CATS ARE

NOT PEAS

Laura Gould

CATS ARE

NOT PEAS

A Calico History of Genetics

COPERNICUS
AN IMPRINT OF SPRINGER-VERLAG

Published in the United States by Copernicus, an imprint of Springer-Verlag New York, Inc.

Copernicus
Springer-Verlag New York, Inc.
175 Fifth Avenue
New York, NY 10010
USA

Library of Congress Cataloging-in-Publication Data

Gould, Laura   (Laura Lehmer)
    Cats are not peas : a calico history of genetics / Laura L. Gould.
        p.   cm.
    Includes bibliographical references and index.
    ISBN 0-387-94796-5 (hardcover : alk. paper)
    1. Calico cats.   2. Cats—Genetics.   3. Calico cats—Anecdotes.
    I. Title.
    SF449.C34G68   1996
    636.8'22—dc20   96-18692
                                                              CIP

Manufactured in the United States of America.
Printed on acid-free paper.

9 8 7 6 5 4 3 2 1

ISBN 0-387-94796-5   SPIN 10539302

For George, of course

(and Max, too)

and in memory of my father, Derrick Henry Lehmer,
whose provocative question
"How does the Thermos bottle know whether to make things
hotter or colder?"
sent his children down the path of lifelong learning.

Preface

I LIKE TO READ PREFACES (AND FOREWORDS AND PROLOGUES), THOSE lonely, neglected texts, often thumbed past hastily on the way to the core of the tome. I read them carefully for what they can tell me about the origins of objects and ideas, and I am often rewarded, as I was in this 1949 example written by Lady Christabel Aberconway:

> Many people may wonder how I came to make this *Dictionary of Cat Lovers*. Briefly, this is the answer. In the early days of the war I travelled twice a week between London and North Wales, at the best a seven-hour journey. After dark, one read by the light of a torch or a bicycle lamp, precariously perched on a knee or a shoulder.
>
> One evening, while an Alert was sounding, a fellow passenger remarked: "That noise is like the screaming of demon cats in agony."
>
> I found the speaker liked cats. The man sitting opposite then declared he loathed them. The woman beside me said, for her part, she liked them. The little man opposite her said he loved them. I have always loved them. . . .
>
> It occurred to me, sitting in that darkened train, that if I could read about people who had liked cats, and if I could read what they had written about their own cats, perhaps I might discover why those exquisite, fastidious, and sympathetic animals are either warmly loved—or loathed. . . .
>
> The idea, conceived in a half-dream, has been carried out,

and this Dictionary is the result. I confess to feeling almost certain that if at the beginning I had foreseen the years of study and research on which I was embarking, and the depressing times when unknown Memoirs or Letters, arid and unrewarding, seemed to be my only reading, my courage would have failed. Yet now, looking back, my most vivid memories are of intoxicating moments when I discovered, sometimes through a written reference, sometimes through the kindness of a friend, new and lovely works unknown to me until then, even perhaps by name.

How charming and amusing it is: the image of the Baroness on the darkened wartime train; the truly random event that sent her off on a journey that was to last not seven hours but years and years; her excitement and dismay over various discoveries; her persistence and passion for the task; her gratitude for paths to unexpected treasures. I have been there too. I recognize it all.

My journey began not in a darkened train but in a brightly lighted garage where evidence of country mice was all too visible. Cats, we thought, we must get some cats — working cats who would patrol the property and keep the vermin at bay. A small and simple thought, reasonable and straightforward, easily implemented. Who could have foreseen its consequences?

In happy ignorance, we visited the local animal shelter and selected two kittens, George and Max, and the choice of George (based largely on his ability to get along with Max) turned out to be the random event responsible for this lengthy odyssey. For George was a calico cat — a male calico cat — and calicos are invariably female. He was a genetic anomaly, a manifestation of something that isn't supposed to happen, a creature so rare that even most vets have never seen one.

George was also an instigator of infinite questions. My curiosity about his existence has caused me to learn some basic genetics, to examine its history, to explore the calico folklore, and to think about evolutionary change. Thus George has sent me to a multitude of libraries and kept me in my studio on beautiful afternoons, with my nose deep in a book or my eyes squinting wearily before a computer screen; he has caused me to become a bore at dinner parties and a pest on the telephone; he is responsible for all that follows.

In surveying a diverse literature I've been happily surprised by the unexpected humor that has bubbled forth, often unintentionally, from the pages of scientific books and journals that one might expect to be much less effervescent. Consider, for example, the Preface to an 1881 book with the incredibly comprehensive title *The Cat. An Introduction to the Study of Backboned Animals, Especially Mammals*. This sizeable work was written by St. George Mivart, a leading biologist of the time, who felt that change in his field was so rapid that "the Natural History of Animals and Plants needs to be rewritten — the field of Nature being surveyed from a new stand-point." He eschews Man as his "stand-point" because,

> The human body is so large that its dissection is very laborious, and it is a task generally at first unpleasing to those who have no special reason for undertaking it.
> The problem then has been to select as a type for examination and comparison, an animal easily obtained and of convenient size; one belonging to man's class and one not so different from him in structure but that comparisons between it and him (as to limbs and other larger portions of its frame) may readily suggest themselves to the student. Such an animal is the common Cat.

Scientists, of course, are human, and their paths are filled with pitfalls, like those of all the rest of us. Some of their tales are truly remarkable and bizarre. In reading papers from long ago, one is reminded that, given the rapid explosion of knowledge, the next generation of readers may find current efforts just as amusing. We're still stumbling about in semi-darkness as we attempt to make sense of our confusing universe.

Perhaps no one has stumbled more than I, as in my overwhelming ignorance I've attempted both to comprehend a foreign discipline and to describe it in ways that may make it accessible to others. The vocabulary of genetics is both formidable and horrendous. Even its routine parlance is filled with words like *autosomal, blastocyst, epistasis, homozygous,* and so on, *ad nauseum*; its more exotic offerings, like *acatalasemia,* are to be found only in specialized dictionaries of biology or genetics. Although these densely packed terms are efficient vehicles for conveying information to the cognoscenti, they're tough on us newcomers who are having a hard enough time trying to grapple with new concepts. Having been forced to cope with them as a reader, I've tried to avoid them as a writer and have chosen to use "common" rather than "technical" language wherever possible; I've even had the hubris to invent some new and simpler terminology.

Much has been written, of course, about such giants in the field as Darwin and Mendel, Morgan and Sturtevant, Watson and Crick. Here, with the exception of the irresistible Mendel, they're given short shrift, not to minimize their importance but to avoid the duplication of easily accessible material. Except to provide the necessary background, they don't really belong here in any event, for this tale revolves not around peas or fruit flies or DNA, but around calico cats and the people who were curious, as I was, about their origins. The contributions of the cat people (and the cats) to the development of genetics seem to have been largely ignored — a great pity, for their history is both charming and valiant.

Given the conflicting information in what I've read, I've tried, wherever reasonable, to write from primary sources, some quite old and moldy. As one follows the reference trail through the literature, eventually coming full circle and recognizing as old friends documents that once were strange and mysterious, one can see how misinformation is propagated: It's happily copied from place to place, or sometimes miscopied — a mutation of a mutation arising to join the ranks of "facts" to be dealt with.

Among other things, this work represents for me a celebration of innocence. Rather than reading a beginning textbook diligently from cover to cover to acquire the necessary vocabulary and concepts, I made a deliberate effort to learn in a non-standard fashion: to follow my nose and my instincts, to leap into the middle, to learn to swim by almost drowning. This seemingly haphazard approach comes not from laziness but from fear; it represents a need, if you will, to preserve my virginity. Only someone as ignorant as I was in the beginning could have asked the kinds of questions that I did. I was afraid that following the carefully laid out, well-trodden paths of the texts would blunt my curiosity and lull me into believing that I understood things I really didn't.

Having deliberately read in this rather chaotic fashion, I've chosen to write that way as well. Thus rather than peering down from the exalted height of my newly acquired knowledge and mapping out the most efficient route to the pinnacle, I've chosen to let the reader follow in my wondering and wandering footsteps; these have led into more corridors and cul-de-sacs than one would have thought possible. The meandering structure that resulted attests to three beliefs: that a path may be as interesting as its destination, that country lanes have more charm than superhighways, and that there is more than one way to skin a cat.

Over the years my intellectual life, like that of the Baroness, has been most disorderly. It's been characterized by unexpected and lengthy detours caused by random and seemingly innocuous events: an insult in a bookstore which led to years of struggle with ancient Akkadian texts; a chance meeting in a parking lot which propelled me onto the path of machine translation and other computer-based endeavors. So it isn't surprising that George has had such a powerful effect — he's a much more interesting (and more lovable) catalyst than any that has come before.

I've wondered from time to time whether George or I might die before this book was done. (Every answer spawned yet more questions, so that seemed a definite possibility.) George's death, I thought, might be as fatal a blow to this project as would my own. Because George hunts at night while I hunt during the day, our working schedules might seem to be disjoint. Yet he helps me just by curling up on the sofa in a companionable fashion, with his paws tightly covering his eyes. There he sleeps most peacefully in my studio while I struggle and squirm and scratch my head, attempting to fit the pieces together. I seem to need the consolation and encouragement of his daily, albeit unconscious, presence to proceed. Without George I might lose interest, abandon the tangle, and learn to play golf or bridge.

Somehow we've both survived, and I, at least, am considerably wiser for the effort. While George is occupied with "seeing no evil," I see much that is fascinating and am filled with wonder at both the course of scientific discovery and the complexity of even the tiniest creature. I hope that George and I have been able to transmit this wonder, fascination, and humor in the pages that follow.

"Treasure your exceptions!"
—William Bateson

Acknowledgment

I DON'T LIKE TO READ ACKNOWLEDGMENTS, THOSE TEDIOUS, OBLIGATORY recitations of accumulated debt. I avoid reading them because they don't usually tell me anything I want to know. They are filled with the names of people I've never heard of and have no need to pursue: the valiant typist (rapidly being phased out by the author's own encounters with a word processor), the patient and devoted spouse, the neglected children, sometimes even the family dog.

Here my debt is so large that I've decided to declare literary bankruptcy by defaulting on my obligation to mention by name the incredible numbers of diverse participants, without whose bemused assistance I would still be back at square one. Thus the scholars of ancient languages, the many librarians (mad and otherwise), the scientists, the veterinarians, the cat lovers, the folklorists, the various people of Japanese descent, and the huge contingent of helpful friends — all will remain anonymous (no doubt to the relief of many). But I do thank them all, most fervently.

Still, major credit must be given to Serendipity, that lovely lady who continues to rule my life, and to Severo, my remarkable husband, who does his best to rule it when She's not around. He also rules, as best he can, our physical world, which makes it possible for me to retreat so deeply into my mental one. He builds Pooneries, splits wood, struggles with generators, plants gardens and orchards, edits drafts, hugs me tenderly, and plays the piano most beautifully.

In deference to George and Max, we have no dog.

Table of Contents

List of Figures

1
In the Beginning,
There was George...

...and Max too

GEORGE CAME TO US FROM THE HUMANE SOCIETY. AT SIX WEEKS HE WAS tiny and scrawny, embodying the description on his cage, which explained that his "family" had given him up for adoption because they couldn't afford to feed him. He was awkward as well and appeared to be walking on thin white stilts — stilts that were slightly taller in back than in front. He had a long skinny tail, striped like a raccoon's, and a plain white belly. But his back was beautifully colored in patches of orange and black, making me wonder suddenly whether he could really be a male, as his identification card declared. I thought I remembered that all calico cats were female.

However, I wasn't going to ask any questions. The week before, we'd spent several hours falling in love with a charming pair of kittens, only to be told at the adoption counter that we wouldn't be allowed to take them both home because they were of different sexes and might mate before they were old enough for their mandatory neutering. This time we'd decided in advance to select two males. I picked out an elegant, long-haired black tom with a white tuxedo front, whose name seemed obviously to be Max, and at first my husband selected a particularly frisky, feisty coal-black kitten who was great fun to watch in his cage. But when we put them together in a playroom they immediately spat and attacked and tried to scratch each other's eyes out. This was not the sort of relationship we'd had in mind.

We replaced the feisty fighter with the short-haired, long-legged, clumsy calico, who had an amusing kind of charm and a seemingly intelligent face. Now we were treated to a very different kind of encounter: The kittens manifested instant rapport, played happily, and gave each other a bath. So we filled out the adoption forms, stating, despite my suspicions, that the calico's name was George (my husband calls all cats George). We were not to be foiled again.

George and Max spent their first week in the garage, where George kept falling over things and Max almost immediately caught a small frog. (The mice, who had sent us on our journey to the Humane Society, simply packed up and moved out without waiting to see what kind of hunters these cats might turn out to be.) As we spent the first week just watching their antics, I was reminded of an elderly mathematician who had told me years before of his vision of paradise. Imagine, he said, a long corridor, drawn in perspective, stretching back toward a narrow point at infinity. Then imagine that each side of the corridor is lined with straight-backed cane-bottomed chairs. Then imagine that on the seat of every chair there is a kitten.

As George and Max explored the world beyond the garage in the days that followed, some of their antics were reminiscent of old Tom and Jerry cartoons. It had never occurred to me that cartoonists actually draw from life, because their portrayals of action are so extreme. Yet George and Max both went straight up into the air, tails high, feet splayed apart in the prescribed fashion, when they unexpectedly encountered one another coming around a blind corner. And when George leapt clumsily from the level railing of the deck onto the slanting handrail of the stairs below, he had just that expression of horror, and that backward-leaning posture of trying futilely to apply the brakes, that cartoonists portray so vividly. (Fortunately, my husband happened to be standing at the foot of this long flight of stairs and simply scooped him up as he came flying off the end of the rail.)

There were serious things to notice as well. One quiet, sunny afternoon we were playing with George on the lawn when suddenly he streaked away and to hide under the front steps. I hadn't heard or seen anything, but George had noticed the shadow of a hawk and had instinctively run for cover. This reminded us that our kittens were low in the predator chain and would need some protection, at least for a while. It also made us wonder how information about hawk shadows is transmitted and stored.

The vets turn pale

At the end of this blissful week we took them to the vet for their first examination. I watched the vet's lip curl into a knowing smile as we announced that the calico's name was George, and I wondered whether he was going to suggest Georgette or Georgina instead. But as he took a closer look, the blood drained from his face and he said, with considerable excitement, "I've been a vet for twenty-eight years and I've heard that male calicos exist, but I've never seen one. Would you mind if I take him into the back room so the rest of the staff can see him?"

When the vet returned, still looking awed and pale, I expected that he would be able to explain why virtually all calicos are female and how the rare exceptions like George occur. To my surprise, he couldn't. He wasn't even able to suggest where I might look for the answers to these questions. His ignorance, coupled with his palor, piqued my curiosity — there was certainly a story to be unearthed about George, who must be a very rare cat indeed.

As they grew older, we noted that Max was wonderfully graceful and moved with swift, sure instinct, whereas George continued to be clumsy but gave the appearance of thinking: He seemed to employ logical processes; he plotted and planned. He knew that when the sprinklers shut off, a little bubbling fountain would remain just long enough for him to drink from, so when they started their three-minute cycle he would sit and wait. He could discover how to get down from the high storage area of the garage, but Max would remain trapped up there until carefully coaxed and coached. George was definitely smarter. Was this because calicos, lacking males, could not become inbred?

As we shopped around for a vet we really liked, George and Max visited two more clinics during the next few months, completing their initial sets of shots. And the scene was replayed twice more: The vets turned pale but could provide no useful information. The last one murmured in wonder, "There's a penis in there all right" and then mumbled something about XXY, leaving me to make what I could of this cryptic offering. And so the search for George's genetics began.

Why are all calico cats female?
(except George and his ilk)

Although many of the scientific news stories of 1986 (the year the vets turned pale) were about recombinant DNA, genetic engineering, and the finding of markers for various hereditary diseases, I had somehow maintained a profound level of ignorance about these matters. I had vague memories about the nineteenth-century monk Gregor Mendel and his simple but elegant experiments with round and wrinkled peas. I knew that genes, chromosomes, and the double helixes of DNA were their twentieth-century fruition, but I needed a dictionary to discover what these terms meant and how they were related to one another.

Having ascertained that genes were indeed smaller than chromosomes, I went on to larger issues. *Human Genetics*, a freshman text from the local college bookstore, helped to paint some general pictures in my mind, but a more specific understanding of George's genetics came from *The Book of the Cat*, a comprehensive compendium of information lent by an enthusiastic friend. From these joint sources, the following overly simplified picture emerged:

> Chromosomes are thread-like structures found in the nucleus of almost every cell; they are made in part of DNA. They come in matching pairs, one member of the pair providing genetic infor-

mation from the mother, the other from the father. Cats have 19 pairs of chromosomes; people have 23.

Genes are just little pieces of these chromosomes: tiny segments of DNA. Each segment acts as a code and specifies the production of a particular protein. These proteins do a variety of jobs, but most are concerned with keeping things in good running order.

Each gene has a fixed location on its chromosome and helps to specify a certain trait, like blue eyes or orange hair. (Most genes don't have such visible effects, however, because they're largely occupied with housekeeping.)

Each location may provide a choice of different genes that can occur there (one that says yes, give this person Huntington's disease, or one that says no, don't). The choice is often binary, but some locations have a set of three or more related genes associated with them. (For example, you may be of blood group A, B, or O. As an extreme example, cattle have over 600 different genes for blood type, all competing for the use of a single location!)

There's no fixed limit to the number of alternative genes that might occupy a given location, but any specific organism should have only two members of the set to deal with at a time — one on the chromosome from the mother, the other on the matching chromosome from the father.

When the gene from the mother disagrees with the gene from the father, some mechanism must be used for deciding the nature of the offspring. Often a simple choice is made, depending on which gene is dominant and which recessive. (In a disagreement about blue eyes vs. brown, the dominant brown gene always wins out over the recessive blue, at least in humans.) Sometimes the method of conflict resolution is more complicated.

Calico cats arise when the genes controlling orange coat color disagree: The gene from one parent says yes, the hair should be orange; the gene from the other says no, it shouldn't. In this case, for reasons to be explained later, the result is a mosaic — some hairs orange, some black.

The chromosome pairs come in a variety of sizes and shapes, but except for the so-called sex chromosomes, the two members of a pair are always the same size and shape as each other.

Sex chromosomes come in two flavors: X and Y. Mammals with two X chromosomes are female (XX); those with one X and one Y are male (XY).

The two kinds of sex chromosomes differ greatly in size and shape. The X is long, and the Y is very short. Hence there isn't room on the Y for all the genes that fit on the X.

In cats, the gene for orange hair color happens to lie on the X. There's no space for a matching gene on the Y. Therefore, it's not possible for an XY cat (a male) to have one gene saying yes orange and a matching gene saying no orange. So if you see a calico cat, even at a great distance, you can be sure that it's a female. Usually.

Questions, questions, everywhere

At first I felt a flush of triumphant understanding. I could already explain, as the vets couldn't (or wouldn't), why virtually all calicos are female. But then how do you get a George? How could I account for him? The mumbled "XXY" of the third vet gave a clue. Perhaps George had three sex chromosomes where he should have had only two? If so, one of his Xs could say yes orange, the other could say no orange, and the Y could say male. This seemed plausible, but how did it happen? As I reviewed my tiny treasure trove of facts, numerous other questions instantly arose in all directions.

What did Nature have in mind when she left so little space on the Y-chromosome? Why are the sex chromosomes the only pair to be different in size and shape from one another? Isn't this bizarre? What other genes besides those specifying orange hair color do male cats lack because of this real estate problem?

And what about people? What genes do we have on our X-chromosome that aren't represented on the Y? Does this mean that females, with two Xs, have twice as much genetic information about certain traits as males? If so, isn't this an unfair advantage? Does it in some way account for things like baldness, color blindness, muscular dystrophy, and hemophilia, which usually afflict males only?

If there are XXY cats, are there XXY people? Would you know it if you saw one on the street? What about other combinations of X and Y? What determines sex anyway? How did the sex chromosomes get such boring names?

Are XXs females and XYs males in all creatures? What about the birds, where the brilliant colors are characteristic of the males instead of the females? Is it reversed for them? Do the males have extra color genes?

If people have 23 chromosome pairs and cats 19, what about dogs? And mice? And Mendel's peas? Do people have the most? Does it matter?

When did people first notice that almost all calicos are female? How did they explain this strange phenomenon? Did they believe these cats had special properties for good or evil? And what about the rare males? Were they idolized and valued? When did the first calicos — or for that matter, the first cats — appear?

What about current folklore? Do most people know that male calicos are virtually nonexistent? Could they remember how they learned this curious fact?

Do people who own calico cats, especially male ones, understand why this color scheme is usually found in females only? Do they communicate with one another through some society? Is George valuable? Is he an important genetic anomaly? Could he serve some useful scientific purpose? Is he likely to be fertile?

The destruction of data

This last question was of particular significance and urgency because George and Max were now six months old. They had developed a particularly loving relationship and spent hours curled up in an old clothes basket in the laundry room simultaneously washing each other's necks.

They slept front to front with their limbs entwined in a variety of endearing poses, or front to back like two spoons, as Kurt Vonnegut likes to say. Max was bigger and heavier and appeared even more so because of his very long hair. He had developed a protective air toward George, making me wonder whether he perceived him as female; they continued to impersonate the perfect couple.

Besides developing this ideal relationship, they had both developed neat round balls. Max's were a silky black duo, but George's were exotic: one black and one orange, with a neat line between the two. This was troublesome. We had signed a paper at the Humane Society, promising (and paying in advance) to have them castrated before they were seven months old. But if we neutered George, were we destroying a national treasure?

I called a school of veterinary medicine to learn what I could about the prognosis for George's fertility. It wasn't good. I was assured that there was less than one chance in a million that George would have viable sperm. The veterinarian on the phone didn't seem particularly interested in his existence, indicating that everything important about the Georges of this world had already been discovered. She said we should just neuter him, enjoy him for himself, and stop fretting about his uniqueness.

I meant to take a picture of George's beautiful bi-colored balls but failed to have film in the camera at the crucial moment — and then it was too late. The dastardly deeds were done and George and Max resumed their idyllic existence, lazing in their basket or collaboratively hunting rats or snakes (one tracking the head while the other tracked the tail).

Terrors of the night

Since being adopted when six weeks old, George and Max had seen other members of their species only during their three visits to the vets and their rather frightening return visit to the Humane Society where they had been caged next to some fierce feral cats, also awaiting castration. Back home they pretended to hunt one another, stalking and pouncing, sharpening their skills as well as their claws for the wide variety of prey and predators that our isolated country environment provided. George was still long-legged, awkward, and clumsy, but he had proved he could negotiate trees at a rapid rate when necessary. So with some trepidation we now let them loose at night to be their nocturnal selves, hoping that they would be agile enough to escape the jaws of the foxes, coyotes, bobcats, and mountain lions we knew to inhabit our forests and meadows.

Our feline friends were usually to be found in the morning waiting to greet us when we arose. George would be curled in a tight ball on the doormat of the covered porch, paws over his eyes. Max was more likely to assume a sentry position on the railing with his long tail hanging down like that of a Colobus monkey, but black instead of white. Occasionally Max would worry us by not making an appearance, but then around lunchtime he would come nonchalantly strolling in across the meadow. His thick coat would be filled with burrs, and he would seem very pleased with himself; sometimes he would have a large rabbit swinging by the nape of its neck, so heavy that it was dragging on the ground.

Despite his neutering, Max still wandered widely. We had picked him up like an errant teenager as we arrived home late one night and found him near our neighbor's gate, half a mile from home. George, by contrast, always stayed close at hand, hunting behind the woodshed and leaving us little tokens of affection: usually intestines of various sizes and shapes, but occasionally an entire velvety mole, the whiskered snout of a gopher, the paw of a squirrel, or the foot-long tail of a woodrat.

We had allowed the cats full reign over their wild dominion for several months and had just come to believe in their survival skills when we were awakened at four in the morning by horrible screams. Rushing out onto the deck, shouting and clapping our hands to simulate gun shots, turning on all the outdoor lighting, we hoped to frighten off whatever predator had invaded our peaceful surroundings. Waving a powerful flashlight, we soon found Max arched in classic fashion on the high peak of the garage. He was still very frightened and could not be coaxed down — and George was nowhere to be seen.

With heavy hearts we descended into the dark woods, flashlight in hand, valiantly calling his name. But where should we look in this endless forest? The task seemed hopeless. We were virtually certain that George had been eaten and would never be seen again. Weary and downhearted, we struggled back up the hill wondering what life without George would be like, both for us and for Max. It was inconceivable.

As we were trying to adjust to these grim images, we heard a faint and plaintive meow — it was George, high in a tree apparently unable to get down. Our weariness vanished, ladders were fetched, and both cats were retrieved and incarcerated in the garage. Our world was safe again from we knew not what. We could sleep, at least for the moment.

George victorious

Then one morning in the spring it was George, not Max, who failed to give the morning greeting. Max looked lonely and disconsolate. He lay around and complained; as the afternoon wore on, he demanded ever more attention. It became dark and George still failed to appear. We slept badly and hoped to find him on the doormat in the morning.

But he didn't come and didn't come, and once again we were sure he had been eaten. We took Max for long walks through various favorite haunts, hoping George would smell him and return. We shouted for George, we suffered, we waited and hoped. We saw a pair of teen-age bobcats crossing the meadow and shivered with fear instead of viewing them with our usual excitement and admiration. And Max became morbid as well, plaintive and lethargic and needing so much attention that we experienced all phases of a lunar eclipse as we catered to his piteous cries. Finally, after four nights of lonely misery, we gave up hope and decided that this time George was gone for good. In a stunned and still disbelieving mood, we located a calico kitten (female, of course) to console us all in our bereavement.

But just as Max and I were setting off in the car to interview this calico surrogate, the great shout "**GEORGE IS HOME!**" rang out across the countryside. And there he was, ambling in, looking neither tired nor hurt nor hungry nor particularly glad to see any of us — back from some private adventure whose details we were not to know. Within a few hours he accepted Max's solicitous attentions, and once again our lives resumed their former course. George was victorious; he had made our level of dependence abundantly clear. Thereafter he was to remind us in this way several times a year.

Is George valuable?

George and his anomalous genetics had cast a spell over some of our friends as well. They came bearing gifts of information, some popular, some technical, some general — all welcome additions to our small supply. The first batch came in the form of two clippings from a popular cat lover's magazine and immediately answered the question about value with a resounding "No." Although his value to us had been proved boundless, his value to the world was apparently nil. The magazine agreed with the vet school that George wasn't a national treasure. He was just one of those rare accidents that occasionally occurs.

The article went on to say, however, that male calicos used to be of slight financial value and had been sought, some twenty years or so ago, by researchers, perhaps at the University of Washington, who had been trying to prove that the orange gene was indeed "sex-linked" — that is, that it resided on the X-chromosome. Initially they'd advertised for such cats, but eventually they'd just embarked on a program to breed male calicos themselves!

Although one question was answered, many more were again generated: How could you breed genetic accidents? How much of this article could I trust? Was it all nonsense? Where could I find scientific references to the initial sex-linked gene experiments? Did the first such experiments really take place only twenty years ago? How and where should I begin?

What's a calico anyway?

This simple question seemed a good place to start, and *The Book of the Cat* seemed a good place to look for a definitive answer. Leafing through its handsome illustrations of cats of many breeds, with coats of many colors, I soon discovered that *calico* is not the name of a special breed but merely a descriptor of coloration involving black and orange patches. (I also discovered that the term is derived from *Calicut*, a place in India famous for producing parti-colored cotton fabric.)

A calico cat, the book said, is typically two-thirds white; it has large black and orange patches on its back with white dominating the legs and belly. *Tortoiseshell* is a similar descriptor reserved for cats that have no white at all but are covered entirely with a motley array of black and orange hairs. George, it turns out, falls in between and thus should be called a tortie-and-white. That's because he's only one-third white and has small, intermingled black and orange patches dominating his head and body.

Calico coloration can occur in many breeds — in the domestic short-hair, like George, or in more exotic varieties such as the long-haired Persian and the tailless Manx. The black and orange patches may be bright and showy, as in George's case, or dilute and understated, as in the fancy chestnut and lavender calicos seen in cat shows. *Calico*, however, is also a generic term that covers all these cases, and it will be used in what follows wherever the distinctions are irrelevant.

Calico folklore

Many people don't even know what a calico is, let alone the fact that almost all of them are female. And those who are aware of this curious state of affairs can't remember, as I can't, when or where they first heard of it. It's just one of many thousands of such facts about the world (lots of them no doubt erroneous) that clutter up our minds and seem always to have been with us.

Nevertheless, I always try to ask how the fact was learned when-ever I speak with anyone fortunate enough to have a calico around the house, especially a rare male one. The owner of a George named Clyde knew very well: It had been, she said, a "Ramona story" she had loved as a child. All she could remember was that a little boy waited impatiently for his calico cat to have kittens, and when it didn't, the vet discovered that it was actually a rare male, worth a lot of money. So the little boy and the cat went to New York and became rich and famous and lived happily ever after.

Ramona Beasley is a character invented by Beverly Cleary and familiar to millions of children around the world. She represents a powerful force, and I went to the children's section of the library to read the text in the original. There were about twenty Ramona books on the shelf and another ten or so in the card catalog, but none of them appeared to be about a little boy and his desperate need for kittens. The trail grew cold and seemed to dead-end there.

I was thinking about how to pursue the search as I was also trying to listen to the white-haired poet on my right at dinner that evening. The young photographer on my left had already leapt up from the table half a dozen times to go out on the deck for a smoke; the dinner was long and his addiction to nicotine extreme. Neither of us viewed the other as a promising dinner companion, but finally, between hasty departures and with obvious reluctance, he turned and asked me what I "did." I ducked with some flippant remark about calico cats, thinking that would put an end to it, but to my surprise he showed real interest. His favorite story as a child had been about a little boy and his calico cat who didn't have kittens... .

It was my Ramona story, and now I was interested as well. But he couldn't help. He didn't know anything about any Ramona. That was all he could remember. He was surprised he had been able to remember that much. He explained about the psychiatrist he'd been visiting for years in the hope of unlocking the secrets of his forgotten childhood. Almost everything of interest was lost to him; he didn't know why. He was only thirty-five but couldn't remember anything significant before the teen-age years; it was such a trial, such a mystery. With these words he retreated once again to the deck to console himself with smoke.

Moments later he rushed back, all aglow. "While Mrs. Coverlet Was Away," he announced triumphantly, that was the title. And now he remembered, the little boy's name was Toad! And there was lots of purple glop, but he couldn't remember why. It was all coming back. When he returned to New York, he said, maybe he wouldn't need his psychiatrist any more. Some dam was breaking and who knew what might be released... .

And there it was on the library shelf, exactly the strange title he had remembered. It wasn't a Ramona story at all, it was one of Mary Nash's Mrs. Coverlet stories — another powerful force to be investigated in the original. (One could tell it was a powerful force. This particular volume, from the eleventh printing, was the thirtieth copy in our county's library system. This was information that really got around.) So I read about Toad and Nervous, as the rare male calico was called, and there was indeed lots of purple glop. The story wasn't quite as Clyde's mistress had recalled, but it featured the following essential facts: Calicos were usually female, and males were very rare and extremely valuable. Mrs. Hortense Dextrose-Chesapeake, president of the American Cat Club, had come from New York in her chauffeur-driven limousine to pay Toad $1300 dollars for Nervous and had brought him her pedigreed cat, replete with five newborn kittens, in exchange as well.

Was that the going price in 1958, when the book was written? How many people, indoctrinated as a child by Mrs. Coverlet, now firmly believed that male calicos were not only rare but worth a lot of money?

My next bit of calico folklore was discovered much more mundanely. I'd been studying the section in *The Book of the Cat* on the Japanese Bobtail, a breed that often exhibits widely separated patches of orange and black against a mostly white background; there I found that these elegant tri-colored cats are thought by the Japanese to bring good fortune. From other sources, I learned that the luck may manifest itself in various ways. For example, Japanese homeowners with resident calicos may experience improvement in their finances; Japanese sailors with calico cats on board are likely to be safe from storms; and, I was told by a vet, Japanese whorehouses maintain calicos on the premises to ensure the potency of their clientele. (The rare males are no doubt considered superior to the females at all these tasks.)

The precursors of these three-colored cats apparently arrived in Japan, having come from China or Korea, about a thousand years ago. (We know that because they're documented in the writings of those times; they're also featured in many ancient prints and paintings.) Legend has it that the first cats to step on Japanese soil were black; they were followed by white cats and then by orange ones. And so the calico (or in Japanese the *mi-ke* — pronounced mee-kay and meaning literally "three fur") was born.

2

How Do You Get a George?

First you get a calico...

THE FOREGOING DESCRIPTION OF THE ORIGIN OF CALICOS WASN'T SUF-
ficiently explicit to satisfy our visitors, who often asked, "If there aren't
any males to speak of, how do you get more calicos?"

For starters, *The Book of the Cat* provides detailed pictures of reproductive systems that make it abundantly clear how you get more cats,
even calicos. Generally speaking, it's just what one might expect: sperm,
eggs, heat, hormones, caterwauling. Cats, it seems, are a little different
in that instead of releasing eggs at regular intervals, like people, they
release them on demand when they come into heat (several times a
year) and then are prodded into action by the unpleasantly barbed and
spiny penis of the male (that was a surprise). Also unlike people, they
commonly release from three to six eggs at a time. Because cats in
nature are often solitary, not sitting together over the breakfast table or
going to the movies but coming together only to mate, both ovulation
on demand and multiple egg release help ensure successful propagation of the species.

As with people, once an egg is penetrated by a sperm, no rival sperm
are allowed to enter. But there are usually a few more eggs still waiting
to have sperm knocking at their doors. Thus the various kittens of a litter

may have different fathers, which is a mechanism for ensuring genetic diversity. To make the situation even more productive, a cat may sometimes come into heat and mate while pregnant. This often works out badly for the second set of kittens, which is usually born prematurely at the same time as the first. But sometimes there's a happy ending and the second batch is also born alive and well a few weeks later.

So it would seem that cats have lots of opportunities to produce calico kittens. All that's needed is a fertilized egg that contains the following conflicting genetic information: a gene from one parent specifying an orange-colored coat, and a commensurate gene from the other parent specifying a non-orange-colored coat. These genes controlling orange coat color happen to lie on the X-chromosome, so the necessary conflict can arise only in female kittens: They're the only ones who (should) have two Xs. But since half of all kittens are female, this isn't a very serious restriction.

If the mother is an orange cat, then any non-orange males — black ones, tabbies, black-and-whites like Max — could donate the non-orange gene needed for contrast. Here the multiplicity of possible fathers makes a calico offspring more likely. In any of these cases, if the resulting kitten is female, she's almost certain to exhibit black and orange blotches. (To produce a true calico rather than a tortoiseshell coat, a gene specifying white spotting is also needed. However, this gene can come from either parent and lies on a different chromosome.)

To see graphically how this works, consider the following Punnett Square (named after the man who first drew one but modified slightly to suit our purposes). The standard male ♂ and female ♀ symbols employed are probably familiar, but you may not be aware of their wonderful origins: The ♂ represents the shield and the spear of Mars; the ♀ represents the hand mirror of Venus.

The top box of Figure 1 depicts a non-orange male (in this case a black one) and shows the two kinds of sperm he's capable of providing: one with an X-chromosome bearing a non-orange gene, the other with a Y-chromosome having nothing whatsoever to say about color.

At the left side is an orange female, whose egg cells are necessarily identical as far as orange is concerned: Each contains an X-chromosome bearing an orange gene. (This gene is only one of hundreds at work on each X-chromosome, so its size here is greatly exaggerated.)

In the middle, the results of collisions between these various eggs and sperm are depicted, and it's easy to see that all the female kittens will be calico and all the male kittens will be orange, just like the mother.

This diagram does not imply, of course, that every mating will result in exactly four kittens: two calico females and two orange males. It merely specifies the possibilities and shows the proportions that are

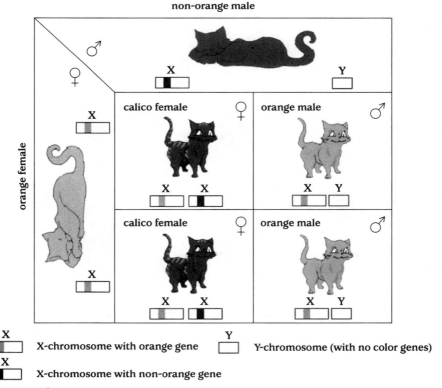

non-orange male

calico female ♀   orange male ♂

calico female ♀   orange male ♂

orange female

X X-chromosome with orange gene   Y Y-chromosome (with no color genes)

X X-chromosome with non-orange gene

Figure 1. Orange female mates with non-orange male.

likely to be observed over the long run. Also specified, beneath each kitten, are the types of egg and sperm cells that they will eventually produce, at least as far as the sex chromosomes and the color orange are concerned.

In Figure 2 the color schemes are held constant but the sexes are reversed: Instead of showing an orange female mating with a non-orange male, it shows an orange male mating with a non-orange female. You might expect, as the early geneticists did, that these reciprocal crosses would produce identical results. But instead there's

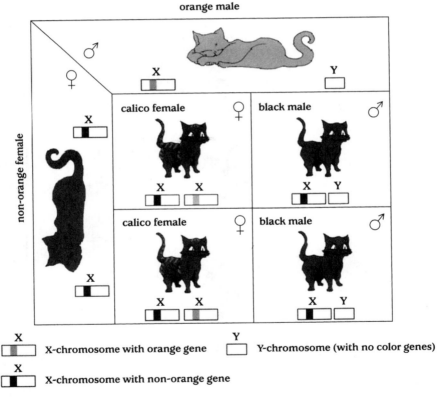

Figure 2.   Orange male mates with non-orange female.

a surprise: Although the female kittens are all calico as before, the male kittens are now all black instead of all orange. (Examination of the respective eggs and sperm will soon dispel the mystery.)

As a final example, consider the mating of a calico female with a non-orange male. This produces a slightly more complicated Punnett Square, with four possible outcomes. The female kittens may be either calico or black; the male kittens may be either orange or black. Each of these four types is equally likely — each has a 25% chance of occurring. There's more variety in Figure 3 because the calico mother has

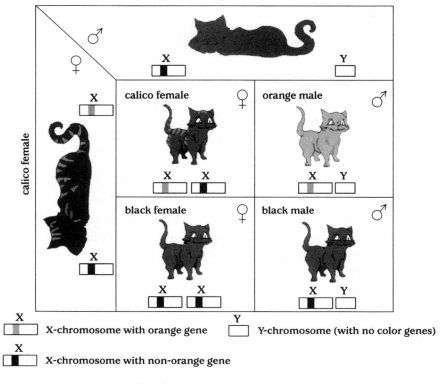

non-orange male

calico female

| | X-chromosome with orange gene | Y-chromosome (with no color genes) |

X-chromosome with non-orange gene

Figure 3.　Calico female mates with non-orange male.

two possible egg types: one with an X that specifies orange, another with an X that specifies non-orange.

Those who feel like pursuing this game further can make their own Punnett Square to see what happens when a calico female mates with an orange male. Yet a different litter type containing a calico will be forthcoming.

All these pictures make it clear that producing new calicos isn't such a mysterious or difficult process after all. In fact, unlike other breeding situations where surprises can result unless the full lineage is known, not much is hidden here. If there's an orange gene around, there's almost always orange hair to show for it.

Studying these diagrams made me wonder why I'd been so concerned about George's fertility. What would he have had to offer that some other cat couldn't provide just as well or better? He doesn't carry a calico gene, or, being only a genetic accident himself, a plan for making more male calicos. He just has a plain old orange gene and will breed (as the early geneticists also discovered to their surprise) just like any ordinary orange tomcat. And there's nothing so special about that.

The mind reels

But how do you *get* a George? Basically, of course, it's just the same old story. Sperm meets egg, and the cells divide happily ever after. To produce a George, however, something must go awry — not too surprising, given that the whole process of making even a new flea is unbelievably complicated. If I was right about what the third vet had meant when he mumbled "XXY" — that the Xs disagreed about the orange and the Y said to make it a male — then George had somehow been blessed with three sex chromosomes instead of the usual two. Plausible, but how do such things happen?

The facts of George's ancestry are lost forever in the anonymity of an animal adoption center. All we really know is that his parents were both cats. Let's assume that they were both run-of-the-mill cats with the standard 19 pairs of chromosomes apiece. Copies of these 38 chromosomes inhabited virtually every cell of their bodies, and new copies were being made all the time, whenever a cell felt the need to divide.

The chromosomes, as you may remember, are composed of simple but seemingly endless sequences of DNA that must all be copied with complete fidelity on every cell division or serious troubles may arise — cancer, for example. An adult human like myself, I learned to my horror, contains about 100 trillion cells, about 1000 times the number of stars in our galaxy. This is a terrifying thought, especially for one who has never been comfortable contemplating the immensities of astronomy. And, I went on to read, they're dividing at a rate of about 25 million every second. That's over 2 trillion divisions every day! So as I sit here musing pleasantly about what's for lunch, I'm also now worrying about whether all my cells are dividing properly. Just think of the opportunities for error!

Having designed computer systems, I was familiar with the importance of redundancy, but nature's design decision seemed extreme. All the genetic information needed by the whole organism, to be supplied to virtually every cell of the body? The skin cells, in effect, containing information not only about what color they should be, but also about what color the eyes and hair should be, and whether or not the organism as a whole will acquire Huntington's disease if it lives long enough?

On further reflection I realized there wasn't a lot of choice. After all, each organism starts from a single cell, which has to contain all the information it needs to get it where it's trying to go. Nature has been at it for eons — it's been about two hundred million years since the first mammal trod the earth and about three billion years since the first living organism appeared — and things seem to work out pretty well most of the time. Still, to quote Vonnegut again, the mind reels.

My mind continued to reel - and yours may too - as I struggled to comprehend, and then describe, the two basic mechanisms by which the cells divide. The first is called *mitosis* (from the Greek mean-

ing '"thread" — an allusion to the thread-like nature of the chromosomes). Mitosis is a basic process that enables a cell to replicate itself
by dividing into two new cells. It's employed by all higher organisms.

The second, more complicated mechanism, is called *meiosis* (from
the Greek meaning "to reduce" — an allusion to the fact that meiosis is a reduction division that cuts the number of chromosomes in
half). Meiosis is a special process used, in mammals, only for the production of eggs or sperm. Since most species reproduce sexually and
hence have need for these products, meiosis is employed by most living organisms.

If you're not familiar with these twin pillars of genetics, as I wasn't
when this odyssey began, please hang on tight for the next few pages
as detailed descriptions are provided. I'll try to make them as painless
as possible, and rewards will definitely be forthcoming.

Mitosis

Mitosis is the process by which a cell replaces itself with two new ones,
identical to each other and to their progenitor. It does this by first dupli-
cating all of its chromosomes — 38 of them in the case of a cat — and
then segregating them, sending one complete set of 38 to its North
Pole, the other to its South Pole (metaphorically speaking). Once this is
done, it proceeds to split in half along its Equator and to enclose each
chromosome set in a new cell membrane. In this way the original cell is
able to replicate itself exactly, providing each of the two new cells that
are formed with a complete and identical set of chromosomes. Very
simple in description if not in execution.

To look at the process of mitosis in detail, consider the 2 pairs of
chromosomes shown in Figure 4 and imagine 19 — think cats. These
pairs are all different sizes and shapes, but the members of each pair
always have the same size and shape (unless they happen to be the
unmatched pair of sex chromosomes, X and Y, which we won't try to
indicate here). Black ones signify those inherited from the father;
white ones signify those inherited from the mother. The floating-free

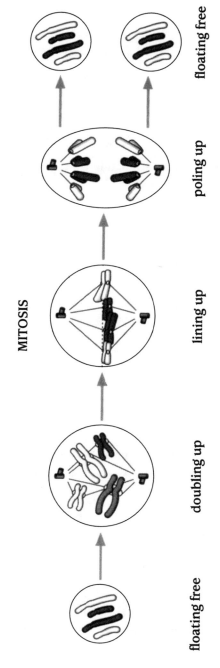

MITOSIS

floating free   doubling up   lining up   poling up   floating free

Figure 4. Making two identical cells by mitosis (only two chromosome pairs shown).

phase that is symbolized in this diagram must be imagined as well, because it's very hard to see: The chromosomes are hiding in a dense tangle in the center of the cell, defying all but the most powerful microscopes.

There, in the dark privacy of the cell's nucleus, each chromosome begins to double itself, becoming a pair of identical *chromatids* connected to each other at a belly-button-like place called a *centromere*. These chromatid pairs are referred to as *dyads*.

Next *spindle fibers* appear, stretching from pole to pole across the cell, sort of like the seams on a football. The centromere of each dyad attaches itself to an available spindle fiber (rope tow) and is pulled back and forth, being slightly attracted first to one pole, then to the other. Eventually equilibrium is attained, and the dyads are lined up neatly along a sort of equatorial plate.

In the poling-up phase that follows, the spindle fibers begin a serious tug of war centered at the poles. They pull each dyad apart, splitting its centromere and drawing its dangling, seemingly reluctant, chromatids to the opposite poles of the cell. These chromatids are identical to the chromosomes with which the cell began. There are now two complete sets of them, one clustered around each pole.

Finally the splitting-up phase occurs, in which the cell begins to divide into two parts. Two new membranes form around two new nuclei, each containing 38 chromosomes (19 pairs), once again floating free. Two identical cells now exist where there was only one before, and the whole cell cycle is ready to happen all over again. Mitosis is over and we're back to square one.

So how long did all this take? Probably a lot longer than it took you to read the simplified account presented here. All this doubling up, lining up, poling up, and splitting up takes a minimum of ten minutes. More often it takes an hour or two, depending on the species, cell type, temperature, and other factors. The doubling-up phase is by far the longest, taking about 60% of the time. This isn't surprising, because that's when all the careful copying occurs.

Meiosis

Meiosis in mammals is the process by which sex cells are generated. It's used exclusively by special cells of the gonads called *germ cells* that have been entrusted with the task of converting themselves into either eggs or sperm. These resulting sex cells are responsible for the propagation of the species and are fundamentally different from all other cells of the body.

One major difference is that each sex cell has only one complete set of chromosomes (all other cells have two). For a cat this means that each egg or sperm cell should contain only a single set of 19 chromosomes, not a double set of 38. The reason for this difference, of course, is so that when two sex cells (of opposite sexes, of course) join up to produce the single cell that will become a new cat, it will acquire the 38 chromosomes needed to do the job properly, not 76.

Another major difference is that the single set of chromosomes in any given sex cell is likely to be different from the single set in any other sex cell (in all other cells of the body, the chromosome sets are usually identical to one another). The reason for this difference has to do with the importance of genetic diversity. Without such diversity, it would be difficult for species to adapt to changing environments; without adaptation, neither you nor I nor George would be here.

To understand how meiosis is used to produce sex cells with these two important characteristics, visualize a germ cell in the testicles of a tomcat. Like all other cells of his body, it has 19 pairs of chromosomes, one member of each pair inherited from his mother, the other from his father. They're all jumbled up together, floating lazily here and there within the cell's nucleus.

Now imagine the germ cell embarking on meiosis by calling the pairs to order. In meiosis, they line up Noah's ark style, maternal and paternal versions side by side, straddling the Equator. Some of the maternal ones will turn out to be on the south side, some on the north — the choice is random. Now envision the germ cell dividing through the middle, sweeping all those to the north into one cell, all those to the south into another. That will result in two new cells, each containing a single complete set of 19 chromosomes, but each set will be quite different from the other.

The choice of which side of the Equator a maternal or paternal chromosome lines up on is random, so every time this process is performed, a different mixture is likely to occur. In fact, since it constitutes a two-way choice made on each of 19 chromosome pairs, that means that 2 to the 19th power — 2^{19} is over half a million! — different kinds of sperm cells could result.

This number of variations on the theme might seem sufficient to provide for adaptability, but it doesn't actually represent as much genetic diversity as you might at first imagine. That's because the genes are still being inherited in large blocks: In this scheme, all the genes of a given chromosome in a sperm cell would come from either the maternal or the paternal side of the tomcat's family.

To deal with this problem, meiosis adds another wrinkle to its randomizing method. Just before the maternal and paternal members of each chromosome pair line up along the Equator, they intermingle in a romantic event called "crossing over." When they enter the clinch, one member of each pair has the genetic information from the tomcat's mother, the other from the tomcat's father, all neatly separated out. When they leave the clinch, however, their situation is quite different. The strength of their embrace often results in some exchange of commensurate parts. Their passion may be such that pieces of limbs break off at cross-over points and reattach themselves to the same position on the matching member. Thus at the end of the crossing-over phase, each member of the pair should contain a mixture of genes from both sources — except when the pair are sex chromosomes.

It's ironic that the sex chromosomes, which one might envision as mixing it up even more than the others, seldom mix it up at all. The tomcat's sex chromosomes don't constitute a matching pair by any means. The runty little Y has only a few genes at its tip that are commensurate with those on the much larger X, so the two of them may just wistfully intertwine their fingertips and otherwise leave each other alone.

That, roughly, is how meiosis goes about its important business of trying to make all the tomcat's sperm cells different from one another: First it produces random mixtures of genes within single chromosomes by allowing matching pairs to intermingle and cross over; then it randomly distributes the now patchwork members of each pair into two new cells as we have described. Since different types of patchwork are likely to be produced whenever such intermingling occurs, it's now possible to make even more than 2 to the 19th different varieties of sperm — and to make much more diverse varieties than before.

The actual phases of meiosis, unfortunately, are more complicated than should be necessary and are undoubtedly the source of many George-like peculiarities. Because meiosis evolved from the earlier and much simpler process of mitosis, it starts out in a copycat fashion by doubling all its chromosomes, turning them into dyads. Since the goal is to cut the number of chromosomes in half, doubling them to start with doesn't seem like a sensible way to go. It isn't, and meiosis has to pay for its folly by then performing two divisions instead of only one.

The first division produces two cells as described for mitosis. Each of these two cells contains only one element of each chromosome type, but the elements are now dyads instead of single chromosomes because of the initial doubling that has occurred, so each cell has to be divided again. By means of these two divisions, four sperm cells are produced, each with the requisite single set of 19 chromosomes on board, ready for action.

Those who wish to understand the process of meiosis in all its gory detail should look carefully at Figure 5. Those who don't will be relieved to learn that only a general understanding of this reduction division is necessary in order to comprehend what follows.

Although the general mechanisms described here apply equally well to eggs and sperm, these two types of sex cells have quite different characteristics. The egg is the largest cell of the body, the sperm the

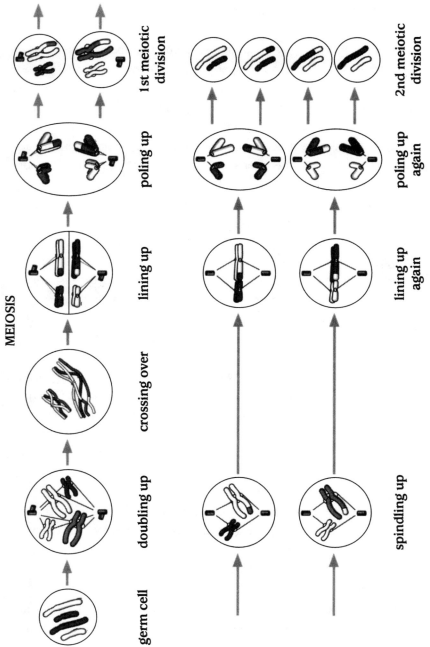

Figure 5. Making four non-identical sperm cells by meiosis (only two chromosome pairs shown).

smallest. For humans, over a quarter of a million sperm could fit inside a single egg if they were allowed to. Eggs are so large because they contain food for nurturing the sperm; the sperm are lean and mean, rushing around looking for an egg to take care of them.

The time they take to complete the meiotic process is also dramatically different. Sperm are being manufactured constantly, on demand, in a process that takes only slightly longer than mitosis. By contrast, a mammalian egg completes the final phase of meiosis only when it's fertilized by a sperm. For some human eggs, this can take as long as fifty years!

These basic differences in size and function make it necessary for the production methods of these cells to be somewhat different as well. Thus the symbolic diagrams of meiosis shown in Figure 5 are applicable to sperm but not to eggs: A single sperm-producing germ cell will indeed be transformed into four same-sized sperm, but a single egg-producing germ cell will be transformed into one large egg and three small "polar bodies," in the manner to be described momentarily.

According to Irene Elia in her book *The Female Animal*, human embryos may start out with millions of cells that have the potential to become eggs (although hundreds of thousands of these may degenerate during the gestation period). By the time a female embryo has completed three months of fetal development, her germ cells have already entered the initial phases of meiosis; at birth they will number about two million. These germ cells will remain in their arrested state until the child-bearing years arrive. Then hormones will cause only a few hundred of them to resume their meiotic progression. Every month one of these will be ovulated, just after it completes its first meiotic division.

Where eggs are concerned, the two cells that result from this division are quite different in size from one another (rather than the same, as symbolized for sperm in Figure 5). One, the *polar body*, is small and is allowed to degenerate, but it may split into two more polar bodies before doing so. The other is large, because it has lots of food on board, and is sent on its journey toward the uterus. If it's penetrated by a sperm during its twenty-four hours of susceptibility, the second meiotic division can finally occur. This second division also results in two products: another degenerating polar body and a single cell in which the egg and sperm are fused.

This single compound cell is called a *zygote* — from the Greek meaning "yoked" (not to be confused with "yolked"). It's the only cell that is formed from the union of two other cells, rather than from the division of a single cell. It's also a very important cell because it contains all the genetic information needed to make a new organism — like you or me.

What's in a name?

Mitosis, meiosis, chromatid, centromere, dyad, and now zygote. Egad. What strange terminology will show up next? Does it matter? What's in a name?

Lots. Names make pictures in our minds, enable us to remember objects and concepts, and connect them in our complicated mental networks to other objects and concepts with which we're already familiar.

The famous British zoologist William Bateson is responsible for much of the standard terminology of genetics. By 1901 he'd already coined a fair number of important terms, and in 1906 he finally established a name for this new field by writing, "To avoid further periphrasis, then, let us say Genetics." Bateson died in 1927, but as the German geneticist Renner reminds us in 1961, "Wherever geneticists are assembled, Bateson is among them — in their technical language."

The word *gene*, however, was coined by Professor Johannsen of the University of Copenhagen, who suggested in 1909 that Darwin's term *pangene* be shortened. In reflecting on this word, and on others that he proceeded to add to the collection still in use today, Johannsen writes in 1911

It is a well-established fact that language is not only our servant, when we wish to express — or even to conceal — our thoughts, but that it may also be our master, overpowering us by means of the notions attached to the current words. This fact is the reason why it is desirable to create a new terminology in all cases where new or revised conceptions are being developed. Old terms are mostly compromised by their application in antiquated or erroneous theories and systems, from which they carry splinters of inadequate ideas, not always harmless to the developing insight.

Predictably, the new terminology of genetics came to consist of long words like *heteropycnotic, phenotype, euchromatin, tetraploidy,* and other polysyllabic inventions, mostly Greek in origin. Although these may have been carefully chosen and pertinent to the task at hand, they certainly make no pictures in the minds of those encountering them for the first time; one might even suspect that they were invented to keep the likes of us at bay. (The term *allelomorph,* chosen by Bateson in 1901 to describe a member of a set of genes that are all associated with a particular location, is a prime example. It's since been shortened to *allele,* but without much benefit to the novice reader. When my granddaughter first encountered *allele* in her high school biology text, she thought it rhymed with ukelele and could never remember what it stood for.)

And even a nice new word like *gene* — short and sweet, easy to pronounce, and well defined by its originator — is now fraught with difficulties. By 1952, two British geneticists felt obliged to decry in print the fact that "the term gene is in current use in at least two distinct ways" and that it had been in this state of disrepair since about 1939. As they point out, it can be used either for "an individual hereditary unit" (as Johannsen intended) or "as a collective term for all the alleles at a locus."

A somewhat similar fate has befallen the term *germ cell,* which was first used in 1855 to mean "the first nucleated cell that appears in the impregnated ovum." By 1868 it appeared in the phrase "sexual distinction of the generative cells into sperm-cells and germ-cells." And now

Webster's Third tells us it may refer either to "an egg or sperm cell, or one of their antecedent cells." Take your choice, and hope that you make the right one based on the context.

And then there's *chromosome complement*, which Webster's also tells us may be either "the entire group of chromosomes in a nucleus" or "the chromosomes received from one parent." Genetics, with its multiple levels of entities, seems particularly prone to ambiguities of this nature. To avoid these problems here, the meanings of such terms have been restricted as indicated in the text and in the very Informal Glossary provided at the back of the book.

Most of the readers of this book are probably not geneticists, so a lot of the standard nomenclature of Bateson and his early co-workers has been abandoned. It's been replaced by the nonstandard use of other, more familiar words in the hope that these may indeed carry splinters of ideas that will be helpful in forming a general picture of what's going on. Thus the Glossary contains some words that look technical and some that don't — but in this context, some that don't, are. The list of such terms has been kept to a minimum, but as new words like *zygote* continue to infiltrate the pages that follow, you may find that a quick peek in the Glossary is refreshing.

Those who wish to pursue genetics more seriously, of course, will have to speak its language. But for them as well as for the rest of us, I hope that my more mundane words will form a bridge to basic understanding and that they will be helpful rather than harmful to the developing insight.

The karyotype of a tomcat

Figures 4 and 5, illustrating mitosis and meiosis, display cells in which only 2 of the 19 pairs of cat chromosomes are visible; these are presented in symbolic form, and you're asked to imagine the remaining 17. By contrast, Figure 6 is an actual picture of all 19 pairs of a tomcat,

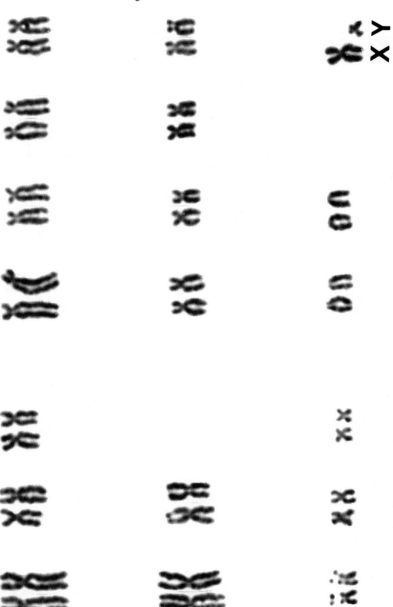

Figure 6. Standard tomcat karyotype, 18 pairs of chromosomes + XY.

caught in their doubling-up phase when they're best visible. These chromosomes have been magnified approximately 5000 times.

Most of the chromosomes, even the runty little Y, look rather like Xs, and I remember thinking, when I first saw such pictures, that they illustrated crossing over — but they don't. The chromosomes look like Xs because they're in their doubled-up form: Each is seen as a dyad, a pair of chromatids tied together at the centromere. The parts extending from the centromere are called arms (although some of them might look more like legs). Thus a gene's location is sometimes described as being on the long or the short arm of a certain chromosome.

Although chromosomes are very experienced at lining up, even doing it differently in meiosis from mitosis, they didn't produce this particular line-up voluntarily. They were coerced into these positions by a technician, armed with both knowledge and scissors, who carefully cut and pasted enlarged pictures of them to produce this standard configuration, which is called a *karyotype* (from the Greek meaning "nut" or "kernel").

A karyotype consists of a full set of chromosomes taken from a single cell. First the pairs are determined; then they're arranged according to various standard conventions, the sex chromosomes always coming last. (The sequence for cats was agreed on at a conference of mammalian geneticists held in San Juan, Puerto Rico, in 1964 and is hence called the San Juan convention.) The pairs are arranged from the longest to the shortest and are grouped together according to the placement of their centromeres. The centromere often occurs in a roughly central position, as its name implies, but it occupies a slightly different location for each chromosome type; for some it is found way at one end or the other, as in the last pair shown before the X and Y.

You've probably noticed that the first two chromosomes in the bottom row seem to be wearing little hats. This effect is from a secondary constriction, rather like a centromere, that causes this particular chromosome type to appear discontinuous. Similar hats can be seen in the karyotypes of lions, jaguars, and ocelots; in fact, there's a remarkable similarity among the karyotypes of all cats, great and small.

Reluctance

Well all right, all right, I can hear you saying, enough about karyotypes and all that. How *do* you get a George? (Patience, patience, we're getting close to the nub now, and all will be revealed momentarily.)

We know how George's parents should have made their sex cells — sperm cells for the tomcat, egg cells for the momcat — containing only 19 chromosomes each. But things don't always go quite according to plan. Occasionally a matching pair of chromosomes finds itself quite reluctant to part — a situation termed (technically and unromantically) *non-disjunction*. Since this doubly negative word is hard to remember and probably evokes no images in anyone's mind, I've decided to use the term *reluctance* instead, at least to start with.

Any of the chromosome pairs may manifest reluctance, and the result is usually disastrous: In humans, a Down syndrome child may be produced (in the case of an extra chromosome 21), or the progeny may be nonviable. Reluctance may occur during the first or the second division of meiois, or in the single division of mitosis when the doubled chromosomes are being split apart and returned to their singular form. This can really screw things up as well. But if it's the sex chromosomes that exhibit reluctance, at any of these partings of the ways, things are not necessarily so bad as they might be.

Remember that female mammals have two X-chromosomes; males have an X and a Y. So when egg cells are formed, the female's pair of Xs are split apart, leaving only one X in each egg cell (that's how a calico mother may pass along either an orange or a non-orange gene to her kitten). When sperm cells are formed, the male's XY pair is split apart, leaving either one X or one Y in each sperm cell. That's why in mammals it's the male that determines sex: If the sperm that penetrates the egg bears an X, the resulting embryo will be female; if it bears a Y, the result will be male.

Thus one answer to the question "How do you get a George?" is that either his mother or his father had a pair of sex chromosomes that were so reluctant to part that they didn't: Either his mother's egg provided him with two Xs that were then joined by a Y from his father's sperm, or his father's sperm provided him with both an X and a Y that were then joined by an X from his mother's egg. These alternate scenarios are shown in Figure 7.

Either circumstance will result in an XXY zygote, a single-celled fertilized egg that will then proceed to divide and divide, making mitotically more and more identical cells, all identically "wrong." And either way the product is a George, a male XXY cat, a cat with 3 sex chromosomes where he should have only 2, with 39 chromosomes instead of 38 in almost every cell of his body — a super cat, a special cat, our own peculiar George.

Perhaps. But are there other ways to account for his existence? How can I find out what his chromosomes actually look like? How is the test done? Does it hurt? Will it cost a lot of money?

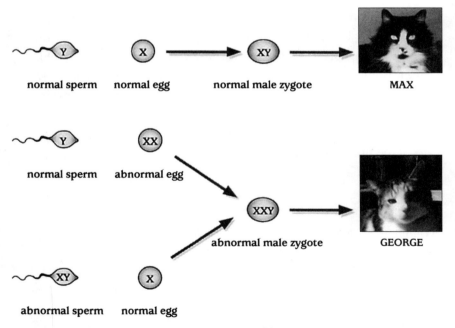

normal sperm   **normal egg**   **normal male zygote**   MAX

normal sperm   **abnormal egg**

                                            abnormal male zygote   GEORGE

abnormal sperm   **normal egg**

Figure 7.   Two ways of making an XXY George by reluctance.

The mad librarian

All the information about why there aren't any male calicos to speak of, how you get more calicos under such circumstances, and how the occasional Georges occur had been gleaned fairly easily, using only *The Book of the Cat* and *Human Genetics* for reference. But the remaining questions seemed harder, and there were more of them all the time — the search was getting serious.

Armed with my alumna card, I drove the considerable distance to my alma mater, hoping to find in its extensive biology library some articles about XXY people, the imbalance between the X and the Y, which species had the most chromosomes, and lord knows what other bizarre facts in this strange and unfamiliar territory. I'd been warned that this was likely to be a sobering experience. Parking would be impossible, the place would be swarming with students absorbing all the resources of both the librarians and the computer systems, half of the computers would be down anyway, I wouldn't be able to find what I needed in a field with which I was unfamiliar, and people would treat me as if I didn't belong (which I didn't).

Instead everything went swimmingly. Parking was easy, the students were elsewhere, and computers were not only available but also accessible to a novice, immediately providing information about whatever I wanted to know. I began by typing "sex chromosomes" and instantly found a number of pointers to books in the open stacks nearby. Then I investigated "cat genetics," and one reference provided a real surprise: The innocuous-sounding *Genetics for Cat Breeders* was listed as being in Locked Case Number 2 along with the books on aphrodisiacs — apparently people tended to steal it! But that aside, I spent several happy hours delving into mysterious tomes on cytogenetics with incomprehensible vocabularies and pictures taken via electron microscopes. It was indeed a strange new world having no relation to that of the very dead languages in which I had immersed myself at this university many years before.

The librarian was happy to accept my alumna card and explained that the books were due in a month but could be renewed by phone if no one else wanted them. The particular books I had chosen had not been checked out for years, so it was unlikely that there would be much contention for their use. I returned to the country triumphant, excited by a reference I had found to human XXYs, who were said to be suffering from Klinefelter syndrome and were characterized as having disproportionately long legs, just like George! I had also come upon an evolutionary theory about what had caused the Y to shrink (apparently, like all the other chromosomes, it used to be the same size as its friend the X). This fit in well with a theory in another book suggesting that all mammals are basically female and that the sole function of the Y is to override this predilection. I was starting to believe that I might be able to manage both the geographical and the intellectual distances involved. I was hot on the trail of many foxes.

My second foray to the biology library bore no resemblance to the first. It came after a long vacation, at the start of which I had returned all my books to the library via a friend. The book about the evolution of the short Y had been particularly interesting and particularly difficult, and I now wished to retrieve it for further perusal. I was brimming with confidence, knowing exactly what I wanted, where it could be found, and how to check it out — nothing could be simpler. Or so I thought until I encountered the rather supercilious young man who this time was in charge at the check-out desk.

"This isn't a library card," he said, looking with disdain at my alumna card. "You can't use this. You'll have to go to the main library and get them to issue you a proper library card." My feeble protest that it had worked last time only made him more adamant, and the next hour was spent trudging up to the main library in the heat and standing in a very long line. Armed at last with the right equipment, I returned to the desk to retrieve my book.

"You can't check this out. This is part of a monograph series and isn't allowed to circulate." This time my protest that I had not only checked it out before but renewed it over the telephone really made him angry.

"We'll see what the computer has to say about this." The computer said fine, check it out, which was really the last straw.

"All right, all right, you can check it out, but only for one day," and he stamped tomorrow's date in the back with real vengeance. I explained that I lived far away and that a single day was of no use, but there was no way he was going to relent. So with dismay, I handed back my now forbidden fruit and left for the long drive home, thoroughly defeated.

Frustrated, discouraged, and still smarting over my failure, I did what one does under such circumstances: I called my mother. She has a faculty library card, and I asked her to try her luck at releasing my book from bondage. She encountered the same young man, who informed her, very politely, that the book was temporarily misplaced — could she please come back again tomorrow? Tomorrow the whereabouts of the book were still unknown, and for months thereafter she received a weekly mailing from a computer informing her that the book was mislaid but a search was underway. Eventually even the computer gave up and sent her a notice declaring the book to be permanently lost. No doubt it had been carefully sequestered from the likes of me by this zealous librarian, diligently protecting his treasures from the gaze of the unworthy.

Fear of pickups

George is afraid of pickup trucks, and with good reason: They come bearing dogs. In the beginning it was Max who showed more caution, keeping his distance and retreating to rooftops, while George's curiosity would get the better of him and he would edge ever closer, sniffing and circling. The dogs were curious too about this small, patchwork creature who showed no fear; they also sniffed and circled while Max watched prudently from on high.

Inevitably a chocolate lab gave chase, completely satisfying George's curiosity about dogs and what they're up to. He barely escaped — by the skin of his tail — and though the bitten place eventually healed, I fear his psyche may be damaged forever. Just the sound of a pickup in the drive now causes George to flatten himself to the ground and slink away in a Groucho Marx-like walk. With his long legs somehow scrunched up beneath his low-slung body, he crawls slowly and silently until he reaches the tall grasses of the meadow. Then he streaks across it, fully extended like a cheetah, making for the safety of the thick woods beyond. Max continues to watch with cautious complacency from some rooftop, but nothing more will be seen of George until the sound of the pickup is heard in the distance, changing gears as it retreats up the steep, narrow road to the ridge.

One spring morning George was having breakfast in the laundry room when a friend, driving the customary pickup replete with dog, arrived to help us fence the orchard. The outside door was shut, but George knew it could be opened at any moment, allowing boundless canine energy to burst in and wreak havoc in his place of refuge. Unable to slink toward the meadow, he somehow slank into the narrow space behind the washing machine. And there he spent the entire day, wedged among the plumbing hoses, facing directly into the corner, making himself as small and invisible as possible. Nothing could coax him out until the last fence post had been set and the frightening vehicle, with its even more frightening cargo, had driven noisily away.

It was sad, we thought, to see the formerly curious George reduced to such behavior. We retrieved a picture of him inspecting a deer at close quarters in the meadow. There'd been plenty of time to take this picture as the encounter had lasted for many minutes, the half-grown kitten and the full-grown doe slowly approaching one another until they were just a few feet apart. Eventually a noise from the woods made the doe skittish and she bounded swiftly away, but George had simply stood his ground, fascinated by this enormous and interesting creature, so different from himself. Just as well, we thought, that he now has this exaggerated fear of dogs. There have been a lot of coyotes in the neighborhood lately.

3
George's Ancient Ancestry

Felis genesis

WHEN WERE THE FIRST CALICOS ANYWAY? OR FOR THAT MATTER, THE FIRST cats? When I originally asked myself these questions, I was thinking about cats like George, for the first part, and Max, for the second. Finding definitive answers to such questions, however, is very difficult. Illustrations of calicos and magpies (as the black cats with white fronts are sometimes called) have appeared only in comparatively recent times, but their history is no doubt much longer than the artistic or written record indicates. It's possible, as noted earlier, that calicos first appeared in the Orient, where beautiful drawings of the *mi-ke* ("three fur") have been found from about A.D. 1000. Where and when they actually originated, however, will probably never be known.

In the several years since these questions were posed, my interests have shifted toward evolutionary theory, and I now take a longer view. So the question "When were the first cats?" has shifted to "When did the first cats of any kind become a new branch of the evolutionary tree?"

The answer is about forty-five million years ago. It seems that this development had to wait for the demise of the dinosaurs, which happened about twenty million years before that. Mammals had already been in existence for well over a hundred million years by the time the largest of the dinosaurs died off, but all that time they had remained quite small — about the size of martens or weasels. They had no chance to become large and dominant, as they are now, with all those dinosaurs around.

The cat family, known as the Felidae, evolved from some early mammalian carnivores called Miacids, which had already developed the intricate mechanism of retractable claws. Given enough time (and no dinosaurs), some really great cats appeared: There was a huge cave lion in Europe, a giant tiger in Northern Asia, and the saber-toothed tiger that roamed all over Europe, Asia, Africa, and North America about thirty-five million years ago. All are now extinct, perhaps because they had the unfortunate combination of a small brain and a large body, instead of the other way around. The saber-toothed tiger was the most successful of these large creatures and terrorized the land for millions of years. It may have roamed our California hills, as the mountain lion does now, as recently as thirteen thousand years ago.

Most zoologists agree that there are 38 species of Felidae, including the domestic cat. These are usually grouped into two main genera on the basis of the ability to roar. The roarers (lions, tigers, panthers) are generally larger, but this is not the determining factor; mountain lions weighing more than a hundred pounds have no ability to roar (although we do hear them scream occasionally during the mating season).

The determining factor is a long skinny bone at the base of the tongue that connects the larynx with the skull. In some cats, parts of this bone are actually not bone but cartilage, which allows the vocal apparatus to move around and reverberate. In others, it's solid bone all the way. So if you've got the cartilage, you can roar and you belong to the genus *Panthera*; if you don't, you can't and you're a *Felis*. (Too bad for George and Max — wouldn't it be marvelous if they could roar! In compensation, they can purr, but surprisingly, no one really knows how they go about it.)

Most members of the family Felidae have 19 pairs of chromosomes, just like George and Max. The exceptions are the ocelots and the Geoffroy's cat, which have only 18 pairs and are sometimes classified as the genus *Leopardus*. Since the cheetah is the only member of the Felidae that can't completely retract its claws, it is assigned to a special genus all its own, *Acinonyx*. (This inability forces it to grip the ground as it runs, and helps to make it - at over 60 mph - the fastest cat around.)

Fossils resembling familiar smaller cats date back about twelve million years, and by the time of the great Ice Ages, only three million years ago, there were already cats of the genus *Felis* around. Reevaluating the question "When were the first cats?" yet again, it might mean "When did the first members of *Felis domestica* appear?" The sad answer is that they came along rather late in the game — perhaps only 4000 years ago. Thus there was a terrible period of over 200,000 years during which humans vaguely resembling us were around but cats vaguely resembling George and Max weren't. Our ancestors, who didn't even know what they were missing, had to make do with dogs, reindeer, bears, and horses. Bones of these creatures dating from very long ago have been found, but no definitive signs of our feline friends exist before about 2000 B.C.

Although *Felis domestica* or *Felis domesticus* were in fashion earlier in this century, I've been told that most zoologists have now reverted to the original nomenclature *Felis catus*, first given to the domestic cat by that famous plant person Linnaeus in 1758. This is not so nice either to read or to write, so I've decided to be one of the holdouts and use *Felis domestica*, that lovely, euphonious designator for the common cat. But *catus, domestica, domesticus* — they're all the same — the species name for a George or a Max.

The luck of the Egyptians

As far as we know, the early Egyptians were the first people to enjoy the company of *Felis domestica* — in fact, they were probably responsible for its domestication. It has been suggested that they succeeded in this task by capturing some of the local African wild cats that hung out around their granaries. When the Egyptians locked them up inside to catch the abundant vermin within, the cats must have thought they'd died and gone to heaven.

That the Egyptians also knew how lucky they were is quite obvious: They worshipped the cat, protected it, and tried to keep it all to themselves. They revered cats not only for their ability to keep the rats out of the grain, but also for their great elegance and beauty. A serious cat cult started up around 1580 B.C. and lasted for almost 2000 years. It revolved around the worship of the goddess Pasht (possibly the origin of the English word *puss*), who had the body of a slender, regal woman but the head of a cat.

During their lives, Egyptian cats were protected by law, and anyone harming a cat did so at risk of his life. After death, the cats were embalmed and mummified before being reverently placed in wooden sarcophagi carved in the shape of a cat. Inside all this protection, the cat mummies were bound in colored bandages and their faces covered with wooden masks that had the features, including the whiskers, carefully painted on. (Also mummified were crocodiles and ibises and even mice, who were entombed with the cats to provide food in the Netherworld.)

Such a lot of cat mummies were found in the late nineteenth century that no one knew what to do with them all. More than 300,000 were unearthed in a single necropolis at Beni Hassan alone. So an enterprising merchant loaded them all onto a ship bound for Liverpool and sold them to English farmers as a cheap but most exotic fertilizer — only $15 a ton.

Fortunately, a few Egyptian cat mummies escaped this ignominious end. These were examined by archeologists and biologists who wanted to determine what kinds of cats these mummies were descended from. They found that most had skulls resembling *Felis libyca*, the above-mentioned African wild cat that patrolled the granaries, but a few had skulls resembling *Felis chaus*, the slightly larger jungle cat that probably hung out there as well. For our purposes, however, we can consider *Felis libyca* to be the cat the Egyptians domesticated, the one from which the branch of *Felis domestica* sprouted on the evolutionary tree.

Unnatural selection

Many members of *Felis libyca* still thrive in the wild in parts of Africa and Asia today. They weigh between 10 and 18 pounds — on the heavy side for a domestic cat — and are a light brown color called *agouti*. (An agouti is a South American rodent, a member of the guinea pig family, which has kindly lent its name to the color of its coat.) Against this agouti background, in which the individual hairs are actually banded in two different shades of nondescript brown, they display slightly darker tabby stripes. Thus they employ several forms of camouflage at once: Their banded agouti coats reduce their visibility by blending well with dry grasses and the ground, while their tabby stripes obscure their shape by simulating shadows and branches. Little evidence exists for evolutionary change in *Felis libyca* over the last 4000 years, so its current members probably look very like their ancient Egyptian ancestors. There's been no need, over this relatively short time span, to come out with a new improved model.

Left alone to mind its own business, evolution is usually a very slow process. Major change occurs at a glacial pace, through many minor perturbations, and one practically needs geologic training to approach the subject with the proper perspective. (In fact, it was the famous geologist Lyell who had the most profound impact on Darwin's early thinking.) Just as I have trouble imagining the immensities of astronomy, I have great difficulty thinking in terms of spans of time covering not just thousands but hundreds of millions of years.

People have tried a variety of approaches in their attempts to reify such quantities, which lie so far outside our normal scales of comprehension. Years ago a friend tried poppy seeds. He kept a jar containing a million of them on his desk to give him some feeling for the number of words in his computer's memory. He would need 180 such jars to represent in poppy seeds the number of years since primitive mammals first acquired one of their chief characteristics: a coat of hair.

The hair protects from heat and cold, and helps to hide its owner from prey and predator alike. Many genes are involved in controlling the hair's color, type, and length, and those that have proved to provide the most successful protection (like the agouti gene) are now shared to some extent by the many different species of mammals that have evolved. Given the oceans of time that nature has had to work on this problem, some genes have become so clever that they now specify changes in the color of the hair depending on the season of the year or the life stage of the organism.

Natural selection is not only slow, it's mindless — it has no goal. The way things work out, however, it eventually and inevitably favors the perpetuation of those genes that help their hosts adapt most successfully to their ever-changing environment. By contrast, the "unnatural selection" of breeding programs is much quicker and is goal-directed: It favors the preservation and perpetuation of those genes most pleasing to people. These genes help their hosts adapt to the "artificial environment" created by the desires of *Homo sapiens*.

The domestication of a wild animal, with the resulting "artificial" production of a new species, is also a slow process, but it does take place during a time span that we can comprehend. For example, people have been shaping the dog to suit their purposes for approximately 10,000 years, and dogs now come in a wide variety of sizes and shapes suitable to the numerous tasks that people hope the dog will help them out with. During this same period of time, however, their wild ancestors the wolves have exhibited little variation.

Similarly, members of *Felis domestica*, which people have been tampering with for only 4000 years, show noticeable changes, while *Felis libyca* has remained more or less the same. Unlike dogs, however, cats are seldom bred for usefulness (cats have strong opinions of their own about how best to spend their lives), so changes in their bodily form are much less dramatic, and no cat versions of St. Bernards or German shepherds are likely to emerge.

But since cats are bred for beauty, changes in the color and type of the coat have been considerable. Natural selection, of course, would have voted against any color that failed to provide adequate camouflage for an animal too small to protect itself, but unnatural selection has changed the rules: In an environment where people-pleasing is of primary importance, the most successful cats are no longer the nondescript ones but are those that really strut their stuff.

Different people, of course, are pleased by different things. There's now a gene that specifies hairlessness (resulting in a very chilly cat called the Sphynx) and one that makes the ears fold down (resulting in a sometimes deaf cat called the Scottish Fold). These cats seem ugly to me rather than beautiful, and I think these variants should perhaps have been abandoned. But who am I to judge? When the first seal point Siamese was displayed in England in 1871, a contemporary journal described it as "an unnatural nightmare kind of cat," but probably no one now would agree with this assessment.

The genes of George
(and Max and Libby)

The cat-worshipping Egyptians left us lots of evidence about what their favorite animal looked like. A particularly spectacular tomb painting, dating from about 1400 B.C., shows a large cat, whom we'll call Libby, catching an almost equally large bird. A splendid mosaic from Pompeii depicts a similar scene and features a cat that bears a striking likeness to the earlier Libby; both closely resemble members of *Felis libyca*.

All these animals are short-haired and of a golden hue. They have elaborate stripes all over their bodies and darkly ringed tails; these striped patterns are what we now call *tabby*. (Like *calico*, the name *tabby* comes from a place where cloth is manufactured. In this case the place is the

Attabiah district of old Baghdad, and the cloth is a striped taffeta called tabbi silk.)

You can tell just by glancing at George and Max that quite a few changes have taken place since Libby was around. Where there was once only an agouti-with-stripes color scheme to choose from, today's cats now come in solid black and white and orange, not to mention the grey and Siamese and exotic smoky hues exhibited by stars of cat shows. And instead of being only short, the hair can now be long — so long sometimes that cats have trouble keeping themselves clean without human assistance. Where did all these variations come from?

The answer, in part, is from division errors during the complicated process of meiosis. Over the course of 4000 years, zillions of mutations must have arisen, since it's estimated that approximately one genetic error occurs in every million sex cells that are formed. A great many of these mutant genes are lost, of course, because most sex cells never amount to anything. For those that do, their mutant genes are often lost as well, unless they turn out to be to the animal's advantage or come to the attention of a breeder. Some genetic accidents are happy, some not — in fact, some are so unhappy that they're lethal. Dominant lethal genes present no problem since no offspring survive to perpetuate them, but harmful recessive genes persist in the gene pool, killing or crippling those offspring that have the bad luck to inherit two of them (one from each parent).

All members of *Felis libyca* have short hair, so there must have been a "wild-type" gene (probably with a very ancient history) that produced short hair. This gene was passed to *Felis domestica*, and short hair remained ubiquitous among domestic cats until a copying error occurred (apparently somewhere in southern Russia). This error resulted in a mutant gene that specified long hair. Like many mutants, this gene was recessive to the wild type from which it originated, so by itself it wasn't powerful enough to make its wishes (or even its presence) known. But when two cats mated, each passing this mutant gene to their offspring, then long hair like Max's suddenly appeared and spread down from

Russia into Turkey and Iran, where the Angoras and Persians come from. Long hair may have conferred some natural benefits (especially in a Russian climate), but no doubt people strongly influenced the preservation of this trait by feeding and sheltering (and probably combing) the beautiful new long-haired creatures that had evolved.

In a similar fashion, mutations of the genes controlling hair color have given rise to the orange gene and the white spotting gene, both of which happen to be dominant to the wild type genes from which they sprang. We don't know exactly when or where this happened, but by studying the distribution of coat colors currently available, we can guess that these may be among the oldest of the color mutations since they are the most widespread. People must have had a hand in the preservation of these genes, preferring to have cats of many colors. The gene for white spotting, manifested in both George and Max, seems particularly designed to be people-pleasing: It often creates an appealing white blaze across the face, a white bib, and dappled paws.

All members of *Felis libyca* are tabbies, and so was Libby and her look-alike from Pompeii. The basic wild-type tabby gene produces a mackerel striped tabby, but two mutants have arisen: a dominant one that produces an Abyssinian tabby (which looks more ticked than striped), and a recessive one that produces a blotched or "classic" tabby. So all members of *Felis domestica* have two tabby genes and are also tabbies, whether they're willing to admit it or not.

Looking at George, one can see the beautiful striped tabby markings all over the non-white parts of his body and especially around his eyes and on his darkly ringed, raccoon-like tail. He proudly exhibits his wild-type tabby gene, which is probably just like Libby's (and like that of all current members of *Felis libyca*). Looking at Max, however, who appears to be solid black (except where he's solid white), one is hard pressed to understand what his tabby genes are up to.

We know the stripes are there, however, because solid black kittens often display the "ghost patterns" of their tabbiness until their hair becomes fully pigmented. These markings can sometimes be seen on fully grown black cats as well, as Kipling reminds us poetically in his *Jungle Book*: "It was Bagheera the Black Panther, inky black all over, but with the panther markings showing up in certain lights like the pattern of watered silk."

If I'd known then what I know now, I would have looked for tabby stripes on Max when he was very young. But now his black stripes are exhibited against a solid black background, making them virtually impossible to see. So there's no way to find out which of the three types of tabby genes Max has; we know for sure, though, that he's really a tabby at heart.

Like Libby, George employs both tabby and agouti camouflage. His black and orange patches are not solid but made up of banded agouti hairs interlaced with stripes — he's a walking advertisement for many genes at work. Since he's both a tortie (tortoiseshell) and a tabby, he should more properly be called a torbie, or actually a torbie-and-white. It's the white, of course, that blows his cover and makes all that camouflage in vain. (Even so, I sometimes have trouble seeing him when he curls up on a stump in our redwood forest, with his long white legs tucked neatly beneath his variegated body.)

By contrast Max is highly visible, even in the dark of night. His luxuriant white bib catches even the starlight and attracts the eye to his every movement. Such nonprovident coloration is permitted by domestication: Most cats no longer hunt for prey but rely on their owners to feed them; most are also safe from predators in their civilized urban environments, which are quite different from those of Libby's time. Even our country cats, which still hunt and are hunted, have indulgent people around to feed and protect them and to rejoice in their extravagant design.

A walk in the woods

Our cats are like dogs — they like to be taken for walks in the woods. We call and they come running, Max obviously eager, with his lush black tail waving like a banner, and George, just as eager but reluctant to admit it. Max likes to be in front, to lead the way, but first he waits to see which one of the many ways we've chosen. Then he dashes recklessly ahead while the more methodical George strolls circumspectly behind, sniffing at scat, listening to bird calls, investigating every scent and sound. We poke along as well, having to stop frequently to pick up the exhausted Max, who lies panting in the path before us. Then we call to the desultory George, who has lagged so far behind that he is now out of view.

All paths lead downward, at least to begin with, and we descend through areas of oak and pine, then Douglas fir and madrone. Going ever deeper, we reach ravines where the coastal fog collects and condenses, providing the perfect environment for the redwoods and their carpets of sorrel and fern. Eventually we reach some favorite objective: perhaps Cat Meadow, where we sit on logs and watch our companions stalk small creatures in the tall grass; or Caretaker's Glen, where they can be counted on to chase one another up the slanting trunk of a large bay tree; or the Big Stump, which they ignore but we always view with wonder, trying to imagine how men equipped only with hand saws and oxen managed to fell and drag off redwood trees of such proportions, almost 150 years ago.

Sometimes, playing lazily at a resting place, George and Max will suddenly become alert, responding simultaneously to some sight or sound of which we are oblivious. Their ears swivel in unison while their pupils pinpoint the same location, but we seldom find out what they're tracking. Once, however, after many minutes of such coordinated concentration, the object of their attention appeared: A fox emerged from the deep woods and made its way silently across a meadow.

After about a mile the complaints begin, but only if we're still descending, still increasing the distance from home. They mew and stage a lie-down strike; they won't proceed without a ride. Once we turn back, however, their exhaustion evaporates. Max gallops up the hill, tail unfurled, leading the homeward charge. But he soon collapses panting and needs to be picked up and petted. George never sprints and drops in this erratic fashion but rather plods steadily along, refusing to be hurried in his meticulous investigations.

Last week I took them through Cat Meadow and beyond — at least I took Max beyond, since George was busy hunting and refused to come. This time my objective was some redwood orchids remembered vaguely from the year before; I was so eager to rediscover their remote and sheltered hiding place that I carried Max most of the considerable distance involved. Returning at last to Cat Meadow, we found no George. Max mewed and I shouted, but to no avail. We stayed and called and fretted as it was getting late. Then we gave up and turned reluctantly homeward. No banners now; Max was dragging and looked tired and sad as we made our way slowly up the old logging road toward the house. I kept calling, but there was no sign of George. I was dragging too and felt worried and guilty, imagining that the lure of orchids had caused the loss of George.

I was surprised and overjoyed to see him, barely visible in the darkening twilight, sitting quietly by the gate at the end of the road. Max and I ran to greet him, but he just waited in haughty grandeur — no similar signs of joy were returned. "Where've you been?" he seemed to be asking with a cross expression on his face. "What kept you? What on earth possessed you to go so far?"

Cat crazy

Lady Aberconway observed (as quoted in the Preface) that cats seem to be either loved or loathed. She meant by individuals at any given time, but it seems also to be true for certain periods of history: Waves of cat loving and cat hating seem to succeed one another, just like other manias.

The first cat-crazy people were the ancient Egyptians, who really were extreme. Not only did they worship, protect, and mummify these animals, they also wanted to be the only cat keepers in the world. But a few cats were smuggled out (or chose to be stowaways) on shipboard, where they were much needed to protect the cargo. Thus they gradually spread wherever sailors plied the seas and were eventually domesti- cated throughout the Middle East, India, China, and Japan. But prob- ably no one has ever loved them as the Egyptians did, who shaved their eyebrows in mourning when a cat died.

The Greeks and Romans didn't really see what all the fuss was about — they had very efficient weasels, martens, and polecats for the protec- tion of their granaries. It isn't even clear whether the Greeks kept cats. An anonymous article on the ancestry of the cat, appearing in the *Jour- nal of Heredity* in 1917, suggests that perhaps the term *ailouros*, used by the Greeks to describe the rat killers they kept on shipboard, actually referred to a white-breasted marten rather than to a cat. (In any event, this term forms the stem of our words *ailurophile* and *ailurophobe*, which mean "cat lover" and "cat hater or fearer," respectively.)

It was from the Romans that the domestic cat spread northward throughout Europe, probably arriving in Britain sometime before the fifth century. Along the way, these Roman imports must have dallied occasionally with the European wild cat, *Felis silvestris*, for they left behind kittens with its darker tabby markings and thickset body type.

One of the earliest written references to cats in Europe occurs in A.D. 936 in the Dimetian Code of Laws designed by the Welsh prince Hywel Dda. Among these laws were a series designed for the protection of cats, assessing their values, and fixing severe penalties for stealing or killing them. Article XXXII says,

> The worth of a cat that is killed or stolen: its head is to be put downwards upon a clean, even floor, with its tail lifted upwards, and thus suspended, whilst wheat is poured about it, until the tip of its tail be covered; and that is to be its worth; if the corn [that is, grain] cannot be had, a milch sheep with her lamb and its wool is its value, if it be a cat which guards the King's barn. The worth of a common cat is four legal pence.

To put this in historical perspective, the famous ailurophile Carl Van Vechten reminds us that a penny at this time "was equal to the value of a lamb, a kid, a goose, or a hen; a cock or a gander was worth twopence; a sheep or a goat fourpence." So even common cats were highly valued in the tenth century. This implies that they must have been pretty new on the scene and not yet had a chance to devalue themselves by multiplication.

It's interesting to note that although many other kinds of animals are referred to in the Bible, no mention of cats can be found. Apparently no cats were on shipboard to protect the foodstuffs of Noah's ark, but Van Vechten provides the following Arabian folktale to explain their genesis: "The pair of mice originally installed on board this boat increased and multiplied to such an extent that life was rendered unbearable for the other occupants, whereupon Noah passed his hand three times over the head of the lioness and she obligingly sneezed forth the cat."

In the early Middle Ages, the official word from the Church was that cats were bad, but people didn't pay too much attention because they knew, like the occupants of the ark, that their lives would be unbearable without them. Religious persecution of cats seems to have started in Europe in the mid-thirteenth century, perhaps because of a revival of popular interest in various pagan fertility cults. Witches were depicted as consorting with black cats for various nefarious purposes, and Nordic witches in particular were said to ride around in chariots pulled by cats and to wear catskin gloves. But this wave of persecution was temporarily stemmed by the plague.

Fortunate were those families who owned a cat during the time of the Black Death, which wiped out between a quarter and a half of Europe's population during the middle of the fourteenth century. Once the Church realized that the plague was caused by the fleas on the rats brought back by the Crusaders from the Holy Land, they decided that maybe cats weren't so evil after all. But by the middle of the fifteenth century, with the plague a thing of the past, cats again ran afoul of devout Christians, who hated and feared them because of their connections with paganism. Medieval Christians were true ailurophobes, and in their time the cats, especially black ones, were not protected but persecuted, sometimes being hanged or burnt alive in bonfires as symbols of heresy. Remnants of these superstitious fears persist today as people continue to change their routes so black cats will not cross their paths.

Cats fared much better in Asia, where the Muslim and Buddhist attitudes were quite different. Mohammed was said to have so loved cats that rather than disturb one by his movements, he simply cut off the part of his robe on which it was sleeping. Buddhists believed that if you became sufficiently enlightened, your soul would enter the body of a cat when you died; when the cat died, your soul would finally be admitted to paradise. In recognition of their importance, sacred cats were kept in temples in Thailand, and a Siamese cat was paraded at the coronation of the king, as a stand-in for his predecessor, as late as 1926.

By A.D. 1000, cats had made their way from China to Japan, where they were kept as pampered pets and not allowed to mix it up with the rats and mice. This situation persisted for hundreds of years, while the vermin population exploded. Mice were particularly attracted by all those silkworms that were busy running Japan's main industry between the thirteenth and fifteenth centuries. Even then, the Japanese were reluctant to allow their precious treasures to be in close contact with such vermin and decided to try intimidation instead. They put paintings and sculptures of cats all over the place, but (not surprisingly) this didn't seem to have much effect, and somehow the real cats got the blame. By the seventeenth century both the grain harvest and the silk industry were in serious trouble, the cats were in disgrace, and the Japanese finally came to their senses. A decree was issued saying that no one could own a cat — they all had to be set loose to go about their harmful, necessary business. And so the grain and the silk were saved.

Meanwhile things must have improved for George and Max's ancestors in Europe; there's a record of a cat show held at the St. Giles Fair in Winchester in 1598. Cats came into favor again when Napoleon's army in Egypt encountered the plague that occurred at the very end of the eighteenth century, but it was the nineteenth-century scientist Louis

Pasteur who finally caused their rehabilitation as a precious household pet. His discoveries about germs made everyone very hygiene-conscious, and there's no animal cleaner than a cat.

For serious attention, however, cats had to wait until the second half of the nineteenth century, when the first Cat Fancy organizations were formed in England. A huge cat show held in London's Crystal Palace in 1871 really started the ball rolling, and the Victorians became cat crazy: Keeping, showing, and breeding cats became a very fashionable thing to do. They tried hard to preserve and enhance the characteristics they fancied, like very long hair and exotic colors; at the same time, they attempted to breed out traits they didn't fancy, like the seemingly ineradicable tabby stripes that the ancient genes of *Felis libyca* went right on providing.

Their endeavors may have caused naturalists, who previously had studied the results of breeding rabbits, mice, and pigeons, to pay attention to cats as well, and they started to record the results of controlled matings. While the cat fancy people were in search of blue ribbons, the budding geneticists were in search of the genes of the cat (although they wouldn't have been able to describe their quest in such terms at the time). Now instead of the medieval clergy, it was the scientists whom cats had to fear. They needed to be especially wary of Mivart, who had suggested (as quoted in the Preface) that it was time for cats in general to step up, pay their dues, and be dissected so people could have a better understanding of mammals.

We're in a cat-loving period at present, at least in the United States, where about 58 million of them currently reside. We now have more cats than dogs, and more money is spent on cat food than on baby food each year. But who knows when another ailurophobic era may dawn, and for what reason? Perhaps the current hysteria over the spread of AIDS will be the (totally unjustified) cause, as vets often describe feline leukemia as an AIDS-like disease of cats. If their best friends turn against them once again, our unnatural pampered pets will really be in trouble.

The Poonery

On Christmas Eve of 1988 it started to snow, very slowly at first and then with increasing persistence, as if the thick white flakes were announcing that this was not just an idle dusting but a real attempt at a White Christmas. So much snow is a relatively rare event at our 2000-foot location within sight of the Pacific, and we go out into the dusk for a snowlight walk.

George and Max follow, as is their wont, and are amazed at what they find. Something has gone wrong with the ground. It seems to stick to their feet and be slippery both at the same time; it's also unusually cold. Max is appalled. He tries to walk with as little contact as possible, his feet leaving the ground so fast that he prances and springs through the air like a startled deer. George, being much more inquisitive by nature, snaps at the falling flakes and tastes them as if he were catching flies. Soon he starts scraping up little balls of snow, which he bats playfully about with his paws.

During the days that follow, the temperature drops to 17 degrees and stays there. The pipes freeze and burst, in their unprotected Californian construction, and threaten to jettison the 8000 gallons of water from our storage tank onto the meadow below, just as soon as the plug of ice that's holding it in place thaws. George investigates the frozen artichoke plants and the long tongues of ice protruding from the hoses, but he then retreats with Max to the warmth and safety of their basket in the laundry room. No doubt they both hope that whatever has gone wrong will soon be put right.

The people, however, have a more rigid and compelling agenda. The project for the Christmas recess is to start construction of a studio in which I can pursue my researches, so my son and granddaughter help us carry what seem like thousands of icy boards down a frozen, slippery slope to the building site at the foot of the meadow. It's so cold that hammering is very difficult and so grey that when the propane truck comes down the narrow, dangerous road to fill our tank, it appears suddenly, like a huge mirage, its yellow lights piercing the swirling mists. I run to the house to gather up the outgoing mail; I hadn't dared to drive the four miles to the mailbox myself and now overdramatize the propane truck as my only link with the outside world.

By the end of the week, somehow the framing is complete. The little cottage with its peaked roof is snuggled under the pines, awaiting its walls and windows. We dub it the Poonery, a diminutive for the name of our homestead, Poon Hill. "What's a poon?" many of our friends wrote, when we sent them our new address some years ago. A few even looked it up in the dictionary and found that it was a large East Indian tree whose seeds are used to produce bitter-tasting medicine and lamp oil. We have no poon trees here, although it might be interesting to try to grow some. Rather, the name comes from an 11,000-foot hill belonging to a Major Poon of the Nepalese Army. On its summit he built a wonderfully rickety tower, with both an up staircase and a down, from which one could enjoy a 360-degree view of astonishing variety. On one side are the snow-capped peaks of the Annapurnas and their lofty friends, all over 22,000 feet high; on the other are the gently undulating hills that lead down to the green and gold plains of India. This latter view is very like the one we have here when the fog has rolled in, concealing the blue Pacific.

The framing done, the people acquire the wisdom of cats and retreat indoors to wait for warmer times. Soon, however, the Poonery is open for business, and since the business has to do with cats, they're invited inside to participate. They take the matter very seriously and make a careful circuit of the single room, sniffing assiduously along the walls and investigating every corner. They claw in a gentle and leisurely fashion at the battered Persian rug that graces the center of the room. Then they choose their respective ends of the sofa on which to recline, bathe, and snooze.

They repeat this ritual to some extent on every entrance. One day George showed up at the door wanting to come in just as I was wanting to go out. He eluded my defensive actions, dashed completely around the room, performed a half-second token clawing of the carpet, and then made a hasty and determined spring for his end of the sofa. There he curled up into a tight ball with his paws covering his eyes and pretended to be instantly asleep. He won, of course, and I settled in to work for another hour or two.

George was to curl up in this fashion for many months, sleeping deeply while I struggled with the mountains of material that continued to accumulate. Would the answers ever be found? Would the questions never end? Why hadn't I been born a cat so I could just relax and enjoy myself?

George meets Oscar

I couldn't relax even at night, since that winter we were much too often awakened by screams and howls, usually at four in the morning. Fearing that coyotes were trying to feast on George and Max, we'd rush unclothed into the nippy night, making as much noise as possible. On the first such occasion, I failed at first to find the black Max, but the beam of my flashlight easily picked up George's long white legs retreating slowly and deliberately down the drive. I called to him to come home, that all was well, but to my surprise he paid no attention and simply proceeded on his way.

We soon came to realize that it wasn't a wild predator that was destroying our sleep, but a formerly tame one — a white-legged tomcat with a brindled coat whose owner had apparently decided he could jolly well fend for himself and had dumped him in our vicinity. It wasn't George I'd seen but Oscar (as he came to be known), who bore a striking resemblance to him. After numerous nocturnal visitations, I still had trouble distinguishing George from Oscar, even though George has orange patches and Oscar doesn't.

Oscar, of course, knew a good thing when he saw it and wanted to move into the neighborhood, but George and Max weren't having any. At first it was Max who took him on while George sat demurely on the steps of the covered porch and watched. Did we have a romantic triangle on our hands, with George at its apex? Thoughts of this sort were soon dismissed, for now it was George who led the fray, while Max sat aloofly on the rail or supervised from the safety of the rooftops. George attacked Oscar with great ferocity, inflicting deep wounds in the back of his neck. George, it turned out, was a mighty warrior.

Although George and Max often hunt collaboratively, each staking out opposite sides of a hole or cornering a frantic rodent between them, we seldom saw them both at work on Oscar. It was as if two against one wasn't fair — it spoiled the fun. One dawn, however, we were awakened by strange sounds and came down from the loft to find George and Max both threateningly arched, emitting low growls instead of their customary high caterwauling. Each held the high ground of a chair, while Oscar had the low ground of the porch between them. Our sudden presence caused Oscar to bolt and George and Max's frozen forms exploded into action. They took off in hot pursuit, George leading Max by a foot or two, and chased Oscar around the decks with a sound like galloping horses.

Oscar leapt off the last deck and streaked toward the safety of the woods with George and Max (and us) not far behind. Then, as we watched in disbelief, Max caught up with George and, in the confusion of the moment, tackled him and threw him to the ground. Oscar, hearing George's squawk of protest as the breath was knocked out of him, stopped dead in his tracks and turned around to see what on earth was going on. He also watched in disbelief while George got up, dusted himself off, and stalked slowly away in disgust. Then Max realized his mistake and the chase was on again.

Not just one but seven mini-snowstorms were to visit us this unusually cold winter. That will drive Oscar down below, we thought, where it's warmer — but it didn't. We were becoming grumpy from lack of sleep. Something had to be done, but what? One night, as I was closing the side door of the garage, I happened upon Max and Oscar about to mix it up within. I grabbed Max, threw him outside, and slammed the door all in a single motion. Oscar was trapped in the larger sense, but how to trap him in the small?

The answer was a squirrel trap, which was very small indeed for a cat of Oscar's size. We baited it with George and Max's favorite food and happily went to bed, knowing that, for a change, we could sleep peacefully until morning. We arose rested and refreshed to find Oscar filling every inch of the available space; he was unhappy and quite subdued by this unexpected restriction of his freedom.

His unhappiness was short-lived, however, as our neighbor decided that Oscar was just the cat for her. She admired his strength and persistence, his ability to fend for himself in the face of all that frozen adversity, his quite striking beauty. So Oscar got his wish and has become a treasured member of the community. He is now truly a Felis domesticus and only occasionally feels the need to engage in a nocturnal bout with George — perhaps because, as far as I know, George always wins.

Ancient Theories of Sex

Animalcules

ONE DAY I WAS BROWSING IN MY FAVORITE USED BOOK STORE WHEN HIGH up, on the topmost shelf, I spied a book by John Farley entitled *Gametes and Spores*. I was interested in mushrooms in those days since Poon Hill had such a wild abundance: red, orange, yellow, blue, purple, brown, and white ones, all waiting, in those years before the drought, to be identified. The gametes of the title were unfamiliar — they turned out to be sex cells — but I knew about spores; I had started to make spore prints from my mushrooms with a view toward eating the ones that passed the safety tests, as soon as I got up the courage.

Having fetched a ladder and inspected the book more closely, however, I discovered that its cover picture was not of spores but of "animalcules," as spermatozoa used so touchingly to be called. The subtitle, "Ideas about Sexual Reproduction 1750–1914," was a lot more interesting than the title; in fact it looked downright promising. So I took the book home for future reference and found that the ideas it described were wilder and more wonderful than I could ever have imagined.

I learned about Antoni van Leeuwenhoek, a seventeenth-century Dutch merchant who liked to play around with biology and optics in his spare time. He had patiently ground hundreds of lenses, producing powerful microscopes capable of magnifying objects 250 times. Thus he was able to look at freshly ejaculated semen (presumably his own) and see, swimming around in it, little worms with globular heads and serpentine tails. He named them animalcules (little animals) in 1677. A million of them, he said in wonder, "would not equal in size a large grain of sand."

Now that the animalcules were visible, a number of people proposed theories about their function. Most thought them merely parasites of the testes, but van Leeuwenhoek and other early viewers of these animated little wigglers saw in them something much more profound: Inside the rounded head they envisioned a completely pre-formed little creature, infinitesimally small, just waiting for the warmth and comfort of a womb to turn it into a growing embryo. Having discovered the sperm, they became confirmed "spermists," viewing the female function as merely nurturing.

Controversies about the extent of male and female contributions to procreation no doubt began whenever humans first started asking questions, and they have raged into our very own century. Early peoples must have been curious not only about how you got more people, but also about how it worked for the plants and the animals. Surrounded by ballooning females of various species, primitive people probably gave females all or most of the credit, at least where mammals were concerned.

But by the time of the Golden Age (as I learned later from another remarkable book, this one a volume by Hans Stubbe entitled *A History of Genetics, from Prehistoric Times to the Rediscovery of Mendel*), the Greeks were starting to wonder about the male contribution of semen, where it came from, what it was good for. Plato and others suggested that it was produced in the brain and the spinal cord. Hippocrates, Anaxagoras, and Democritus broadened this view into the theory they called pangenesis, which held that semen is formed in every part of the body and travels through the blood; they thought of the testicles merely as holding tanks. (This idea persisted into early Christian times, was lost, and then was rediscovered by various Englishmen in the eighteenth century, particularly John Rogers, who was quoted as saying that "semen, as it were the very essence of the body, is chiefly supplied by the brain and carried through innumerable channels to the testicles.")

Many early Greeks believed that females produced semen just as males did, although there was some disagreement about what it was and where it came from. Galen, who was court physician to the Roman emperor Marcus Aurelius, thought female semen was a mucus generated in the oviducts; it was then sent to the uterus to nourish male sperm and help generate the embryo. He describes female semen, however, as a definitely second-class entity, "less abundant, colder, weaker, and more watery" than its male counterpart.

Other early Greeks equated female semen with the menstrual fluid. In any event, the human embryo was thought to result from an admixture of these two flavors of semen. What type of mixture it was and what rules governed the resulting product were matters of contention between scholars. For a short period, female semen was held to be of equal value with male semen in both reproduction and heredity, but by the time of Aristotle the importance of the female contribution was again on the wane.

Aristotle, Diogenes, and perhaps Pythagoras all theorized that semen was a foam formed from the richest blood, produced by an excess of nourishment. Aristotle, however, disagreed with the other physicians of his time by denying the existence of female semen; He declared that females lacked the "vital heat" necessary for its production. He also made a strong case for the distinction between matter and form. In his view, the female provided the matter through her menstrual blood, while the male's semen shaped this matter into the appropriate form and endowed it with energy and motion. Using the analogy of a (male, of course) sculptor, he explains that the female supplies only the stone, while the male does all the interesting and imaginative work of formation. (More than 2000 years later, Darwin was to agree and, for once, put the idea more succinctly: "Woman makes bud, man puts primordial vivifying principle.")

These theories of Hippocrates, Aristotle, and other Greeks of reknown were temporarily lost after Greek civilization disintegrated and the Christians sacked the famous library at Alexandria in A.D. 391. The writings of many Greeks, however, found their way into the Europe of the Middle Ages by a most circuitous route: They were rescued by the Arabs in Alexandria and Asia Minor, translated into Arabic, carried through Mesopotamia and Egypt into Spain, and then translated again into medieval Latin. But the early Christians continued their suppression of some of Aristotle's works for hundreds of years, and it wasn't until 1231 that Pope Gregory IX said it was okay for people to read them once again.

This was just in time for St. Thomas Aquinas (who was born in 1225) to pick them up and run with them. Echoing Aristotle, he speaks of semen derived from blood and surplus food; he also states that since the male semen is in charge of form and motion, its normal course would be to replicate itself as nearly exactly as possible. Thus the birth of females was described as "monstrous," occurring only through some failure, some deviation from the norm. (Since such failures happened about half the time, one can't help wondering what sorts of excuses were offered.)

Even Leonardo da Vinci, who lived into the sixteenth century, mentions Aristotle's views on these matters without criticism, but by the seventeenth century, there was hope once again for recognition of the female contribution. This hope was provided by William Harvey, a professor of anatomy and surgery, who knew not only that fertilization by semen was necessary in all higher animals but also that chicks originated from the yolks of eggs. Taking a giant leap from this relatively small base of knowledge, he boldly wrote a chapter entitled "An Egg Is the Common Origin of All Animals," even though he'd never seen a mammalian egg, and neither had anyone else. He wrote this around 1650, decades before de Graaf was to discover what he thought was the mammalian egg; this, however, turned out to be only the (Graafian) follicle from which the egg emerges.

In the middle of the seventeenth century, Harvey's unsubstantiated egg theories were all the rage, and most educated people were "ovists" who believed that eggs were indeed where all animals came from. Many also believed, especially where people were concerned, that these eggs contained God-given pre-formed embryos. Then van Leeuwenhoek saw the animalcules, and the stage was set for the war between the spermists and the ovists.

Homunculus, homunculus, wherefore art thou homunculus?

What a curious war it was, producing art work of fascinating form. Both spermists and ovists drew wonderful pictures of homunculi (little men) all scrunched up, clasping their knees, hiding one at a time inside a sperm or an egg, depending on which camp you followed. Both camps believed these homunculi were tiny people, perfect in detail, pre-existing miniatures designed by God, just waiting for a chance to grow large enough to survive in the outside world. Each camp gave the other only the most minor credit for its contribution to the process of creation: Spermists admitted that a nice warm womb made a useful incubator and that the egg might provide nourishment for the developing embryo; some ovists believed that the seminal fluid had a beneficial, stimulating effect on the egg, but most held that the sperm were just parasites. (As late as 1868, Darwin still thought the seminal fluid, but not the sperm, was essential to conception.)

The most radical members of both camps believed in a sort of infinite regression: Not only did they envision these tiny pre-formed people, but they envisioned all people ever to be, hiding within the eggs or sperm of the homunculi, and within the eggs or sperm of *their* eggs or sperm, on and on, *ad infinitum* — all formed by God during some initial monumental orgy of creation. (Their ability to postulate these infinitely small human beings may have been influenced by mathematics. Nicholas Malebranche, one of the originators of the pre-existence theory, pointed out in 1673 that "we have evident and mathematical demonstration of the divisibility of matter *in infinitum*, and that is enough to persuade us there may be animals, still less and less than others, *in infinitum*.")

By the beginning of the eighteenth century, it was clear that the ovists were winning the war. Most people of that time believed that nothing existed without a purpose, so the spermists had a hard time explaining why God would want to waste so much of his precious handiwork: all those poor, doomed homunculi, trapped inside their animalcules, with less than one chance in a zillion of ever amounting to anything. For similar reasons the "pollenists," who had arisen in the botanical arena, were also in serious trouble; any fool could see that most pollen was simply wasted, blowing in the wind.

Linnaeus, that most famous of botanists (in his day as well as ours), was definitely not a spermist. He'd peered at semen through a microscope in 1737 and concluded that the sperm "are not in the least animalcules which enjoy voluntary motion, but inert corpuscles which the innate heat keeps afloat, just like oily particles." But he wasn't an ovist either. In fact, he was one of the few holdouts against either camp of pre-existence theory. Although he had written in 1751 (describing a higher level of pre-existence) that "Species are as numerous as there were created different forms in the beginning," he later retracted this view after he and others managed to create new hybrid plants. He still held to a form of creationism, however, writing in 1764, "We may assume that God made one thing before making two, two things before making four; that he first of all created simplicia, and then composita."

Linnaeus was certain from his observations, not only of hybrids but also of mules, mongrels, and mulattos, that both sexes contributed equally to new progeny, but he didn't know how they went about it. Some of his problems arose because he believed that all plants propagated in a sexual manner, when there were all those algae, fungi, mosses, liverworts, and ferns that didn't. He and a few other rebels couldn't really make a convincing case for their equal-contribution theories and were hence discredited.

By the end of the eighteenth century, the egg was omnipotent. Females, after more than 2000 years of being eclipsed by the supposed importance of the male contribution, were firmly back in the saddle again (though relegated, of course, to their important child-producing function). Hardly a spermist had survived the war.

Chip off the old block

Even in very early times, people must have noticed that offspring resembled their parents, to greater and lesser degrees; that hair color and body type and funny noses seemed to be transmitted from generation to generation, sometimes skipping a generation in the process. Pindar, a Greek poet of the fifth century B.C., was a strong proponent of the aristocracy and was convinced that within noble families, masculine virtues were inherited as well as physical features. He was even able to explain away the lazy good-for-nothings who occasionally showed up, claiming that the noble glory was merely "slumbering" and would awaken in later generations.

By the age of Hippocrates, who was born about the time Pindar died, physicians believed that epilepsy, tuberculosis, and melancholia could also be inherited, and from either side of the family. Since in those days equal importance was given to the contributions of male and female semen, which was thought according to the theory of pangenesis to be formed from all parts of the body, they didn't have much difficulty in explaining how inheritance worked:

> The child resembles in more parts that parent who contributes a larger amount [of semen] to the resemblance, and from more parts of the body.

Such Hippocratic ideas, based on equality of participation, were to be happily propagated over the next thousand years.

But Aristotle, who denied the existence of female semen and minimized the female's contribution in general, was in trouble. His "sculptor" theories were great as long as men produced sons in their own likeness, but how to get around it when they didn't? He was forced into a somewhat unconvincing description of a battle between the motion-filled sperm and the restraining menstrual fluid, with the latter sometimes (unfortunately) winning. He punts on many questions of inheritance and opines that

> he who does not resemble his parents is already, in a certain sense, a monstrosity; ... nature has simply departed from the type. Indeed the first departure occurs when the offspring is female rather than male.

Following in Aristotle's footsteps more than 1500 years later, St. Thomas Aquinas is in deep trouble as well. Believing that semen was formed from excess food, he had real difficulty explaining how grandfathers made their sometimes quite obvious contributions — the relevant food was certainly not ingested by them. To deal with this problem, he suggests that there is some sort of "virtue" or "power" in the semen, a quality that emanates from the soul and is passed down through the paternal line of the family. He doesn't bother, of course, to discuss how maternal traits are inherited, viewing all females merely as monstrous mistakes.

During the time of the ovists, of course, matters were just as bad, but in the opposite direction. If the egg was omnipotent and contained a pre-existing embryo, if the sperm was useless and the seminal fluid only stimulating, how then to account for the oft-observed resemblances passed down from the father's side of the family? William Harvey, who had started the egg craze in the first place, had an explanation: He proposed that if a woman's brain can produce an idea in the likeness of an object she perceives, then a woman's uterus should be able to produce a child in the likeness of the man who fertilizes her.

As the ovists claimed the field of battle at the beginning of the nineteenth century, most of them simply ignored this embarrassing question. By now the action had moved from England and France into Germany, where the universities had experienced a rebirth and were flourishing. Research ability, especially laboratory skill, was highly prized, and lots of students bent diligently over microscopes, hoping to see an egg or some other tiny wonder. They were entranced by this new magnified world of the laboratory and never ventured out into the field to see what was really going on in nature.

The mammalian egg was finally seen in 1827, about 175 years after Harvey first postulated its existence; it was found, in those pre-Mivart days, hiding inside the follicle not of a cat but of a dog. By 1842 it was known that the first stage of mammalian development is the division of the egg into two equal parts. Cells had been recognized as the fundamental units of all living things, and using ever more powerful Zeiss microscopes and new staining techniques, budding biologists could even see parts of cells — their walls and nuclei — and watch them divide.

It was also known that sperm are formed not in the brain or the spinal column but in the testes and that eggs are formed similarly in the ovaries. Sperm were no longer given the status of parasites, but whether or how they managed to stimulate the egg into its unilateral action, how many of them were needed to do this (a multiplicity was generally assumed), and whether they actually penetrated the nucleus of the egg in the process were still questions that were hotly debated. Even those who believed in penetration didn't give the sperm its due, relegating it to the role of stimulant or catalyst.

New information from the plant world was keeping pace with that of the animal world, but with more emphasis on observation by the naked eye. Plant hybridizers had gathered masses of confusing, contradictory data, but one thing was obvious to all of them: Pollen had a lot to say about the product, so some fusion of the egg and pollen cells must be occurring — but no one knew how it worked. The egg still reigned supreme while the sperm remained a second-class citizen.

By the middle of the nineteenth century a serious split had developed between the naturalists and the "scientists," who had acquired a superiority complex and tended to view the naturalists as bumbling amateurs. But these field workers could actually see heredity at work and wanted to account for it, whereas the lab workers cared only to discover microscopic mechanisms invisible to the naked eye. No one at the time had the vision to realize that the work of each was important and must supplement the other.

That's where things stood when Gregor Mendel entered the University of Vienna in 1851.

5
The Genesis of Genetics

Just another Moravian monk

LITTLE IS KNOWN ABOUT MENDEL BECAUSE NO ONE PAID MUCH ATTENTION to him during his lifetime. How were they to know he was going to turn out to be the founder of a science with such important consequences? He was just another Moravian monk, mumbling his devotions and messing about in the monastery garden with his many varieties of peas. The locals thought of him with benevolence, but they certainly didn't take his botanical experiments very seriously. The few scientists with whom he attempted to communicate held that opinion as well. By the time the world noticed how wise and clever he was, he had been dead for sixteen years and most of his contemporaries were dead as well. History was in serious trouble.

But lack of information about Mendel hasn't proved much of a deterrent to the many (including me) who have felt compelled to write about him. The elements of the story are so romantic: a poor peasant child, a starving student, the haven of a monastery, years of painstaking research, a brilliant discovery, total lack of recognition, a lost manuscript, an unsung death. As if that weren't enough, there's still the rediscovery and acclamation to write about: the desire of a somewhat guilty scientific community to give posthumous recognition to one of its loneliest pioneers. And after that the slight besmirchment: the gentle suggestion by a statistician that his data were too good to be true. How can one resist?

The details, of course, vary widely in the telling. One account has it that "he had such violent reactions to visiting the sick that the abbot of his monastery found it necessary to relieve him of those duties." Another account, in considerable contrast, reports that in pursuit of his interest in the inheritance of physical characteristics, "he often attended autopsies in a hospital." A coherent picture of the personality is certainly difficult to acquire from such conflicting data, but the general scenario is reasonably clear.

We know that Gregor Mendel was poor and bright and that he wanted to be a teacher. He managed, however, to fail the test for the teaching certificate not once but twice, thus crushing any hope he had of proceeding in that direction. Instead he ended up as abbot of a small monastery in Bohemia and, eventually, as the father of what came to be known as genetics. Failing a test or two doesn't usually bring about such positive consequences, but in Mendel's case those bad grades helped pave the road to greatness.

The cause of his first failure was lack of knowledge in the natural sciences. Young Gregor had been struggling along at a small, second-rate school, trying to make ends meet by doing some tutoring, and was breaking down in all directions. And so it happened that he became a novice at the Augustinian monastery at Brünn, not from any religious conviction but from the need for physical and intellectual support. He received the latter in the form of two years at the prestigious University of Vienna, where he studied mathematics along with a smattering of physics and chemistry, zoology, entomology, and botany.

Although the time was short, his timing was good because the German universities were flowering: They were stressing not just the acquisition of known facts but the cultivation of human individuality and the importance of independent research. This was no doubt crucial to Mendel's future success, but it didn't help him to pass tests. He failed the teaching certificate exam again (partly because of ill health) and retreated once more to the privacy of his monastery, where he had plenty of time to pursue his experiments with hybrid plants, which depended on many generations of data. (In this way he was like Darwin, whose poor health caused him to hide out in the seclusion and quiet of his country estate. Darwin's peace was purchased with personal wealth, Mendel's with religious devotions.)

Fortunately for posterity, Mendel made a careful chronicle of his work, and it is from his own writings, so clear and yet so seldom quoted, that we can learn the most. He completed his now famous paper entitled, with great modesty, "Experiments on Plant Hybrids" in 1865 and presented it at a meeting of the Brünn Natural History Society, near his monastery. Reading it now (in English translation), one is amazed at its total lack of impact at the time. It seems so straightforward, so clearly thought out, so concise. What was Mendel's problem? Did he need an agent? A zippier title? A more evocative first sentence?

Mendel's opening words are curious and make one realize that the pursuit of aesthetics can have unexpected and serious consequences. He begins with the following innocuous remark: "Artificial fertilization undertaken on ornamental plants to obtain new color variants initiated the experiments to be discussed here." Reading this, one imagines an audience with the raptly upturned faces of garden club ladies in a Helen Hokinson cartoon. This mild and seemingly trivial beginning, however, is soon followed by the big picture, which is very big indeed and tells us exactly and succinctly what Mendel had in mind:

Whoever surveys the work in this field will come to the conviction that among the numerous experiments not one has been carried out to an extent or in a manner that would make it possible to determine the number of different forms in which hybrid progeny appear, permit classification of these forms in each generation with certainty, and ascertain their numerical interrelationships.

The first part of this sentence is Mendel's polite way of saying that botany in his day was a pretty sloppy science. Its practitioners grew and cross-pollinated a lot of plants, but they weren't always careful to control unwanted pollination or to write down precisely what they'd been up to. As each new generation of plants appeared, many would-be botanists stood back, scratched their beards, and wrote rather subjective, unrigorous descriptions of what they saw, or hoped they saw, or wanted to see. From these squishy data they then tried to extract firm conclusions.

The second part of the sentence gives an indication of Mendel's quite different bias and approach. Instead of just rushing off to plant things, Mendel probably sat quietly in a corner and put his training in mathematics and the new scientific method to good use: He formed hypotheses about what would happen when he crossed plants to make hybrids; moreoever, he expressed these hypotheses in terms that could be tested numerically. Then he chose what turned out to be the perfect experimental medium — *Pisum sativum*, the common garden pea — and proceeded to implement a carefully designed series of experiments to prove or disprove these hypotheses.

Nowadays, to get such a long-term research project funded, one would write a lengthy proposal to the government, perhaps explaining how the outcome would enable one to make new kinds of nerve gas or colonize outer space. But Mendel used a different method — he asked for the support of his monastery. Eight years and 10,000 plants later, Mendel closed his introduction to the description of his investigations with this deferential sentence:

> Whether the plan by which the individual experiments were set up and carried out was adequate to the assigned task should be decided by a benevolent judgment.

Unfortunately, Mendel never found out just how benevolent that judgment was to be. No one seemed to understand or to be interested in his results, either at the occasion of the reading or during the rest of his life. Whereas Darwin died rich and famous, Mendel died unsung and unknown, just another monk in a quiet backwater, saying his Hail Marys and carefully counting his peas.

Mendel wrote only one other botanical paper, a short description of his experiments with *Hieracium* hybrids (weedy herbs, according to Webster's Third). Here he was much less fortunate with his choice of experimental material, and his results failed to substantiate his earlier success with peas. This was because *Hieracium* knows how to propagate itself asexually, without pollination, and does so whenever it feels like it. Mendel, of course, didn't know this and kept messing around with his forceps and pollen.

He knew something was the matter, however, because he wrote to Carl von Nägeli, a leading botanist of the time, that "With this species I have not as yet been able to neutralize the influence of its own pollen," but he didn't really understand what was going on. He also remarks that "In *Pisum* as well as in other genera I had observed only uniform hybrids and I therefore expected the same of *Hieracium*." But his expectations were not fulfilled, and he was able to write only a slim, unsatisfactory report on these matters. This appeared in 1869 and was entitled "On Hieracium-Hybrids Obtained by Artificial Fertilization," or, in Mendel's words, "Über einige aus künstlicher Befruchtung gewonnenen Hieracium-Bastarde." Bastards indeed — the German falsely seems to reveal his frustrations with this genus.

Mendel's interests extended well beyond the world of plants. He was interested in astronomy and kept meteorological data for many years. He wanted to know how people transmitted their peculiarities from generation to generation, so he made personal observations and also studied the records of the ancient families of Brünn. He wondered whether what he had discovered about peas worked for bees as well, so he kept about fifty hives with different kinds of queens. He studied the color of the bees, how they flew, their general behavior, and their desire to sting. No doubt he carefully recorded these experiments and their results, but unfortunately no such records remain.

In 1868 the monastery finally got around to extracting payment for all those years of support: It elected Mendel abbot. He was kicked upstairs and made into a manager, which proved, of course, to be the death knell for his various researches. By 1871 he was all washed up as a scientist, no doubt drowning in monkish bureaucracy. In 1873 Mendel writes rather pitifully to Nägeli, who had been "helping" by sending him those wretched plants:

> The Hieracia have withered again without my having been able to give them more than a few hurried visits. I am really unhappy about having to neglect my plants and my bees so completely. Since I have a little spare time at present, and since I do not know whether I shall have any next spring, I am sending you today some material from my last experiments in 1870 and 1871.

He died eleven years later, in 1884, having excited no interest in his theories, which were to lie withering like the Hieracia for another sixteen years after his death. Once rediscovered, however, his round and wrinkled peas were to lay the foundation on which modern genetics rests comfortably to this very day.

Inch by inch, row by row

What exactly did Mendel do all those years, pottering about in the garden of his monastery?

Before ever picking up his spade, he familiarized himself with the work of others who were producing hybrids. There was Collandon, a Swiss, who had crossed white and gray mice and noticed that the colors never blended. Mendel himself worked with mice, perhaps to confirm these experiments, before he ever planted a single pea.

Then there was Dzierzon, a priest in neighboring Silesia, who had crossed German and Italian bees and discovered that the hybrid queens produced drones of each sort in equal numbers; from this it followed that a hybrid male would produce two sorts of sperm with equal frequency.

Gärtner, whom Mendel was covertly criticizing in the introductory remarks to his famous paper, had actually done a fair amount of careful work and had observed that "seed taken from an original cross between two pure species produces nothing but seedlings of the same form" and that it would continue to do so more or less forever.

Reflecting on these matters, Mendel came to the conclusion, not popular in his day, that the differing traits he observed in plants and animals were caused by discrete and independent "factors" (he called them *Elemente*) that could be analyzed statistically. This conclusion led him to formulate two important hypotheses, which have since been upgraded and are now referred to as Mendel's first and second laws.

Mendel's first law, known as the principle of segregation, states that every plant (or animal) has a pair of factors for each of its traits — like the pair that specifies orange fur or non-orange, and the pair that calls for blue eyes or brown. When sex cells are formed, these factors segregate so that only one member of each pair finds its way into the seed (egg) or pollen (sperm). (Here Mendel is describing the result of meiosis, a process he had never seen. Mendel would be dead before microscopes showed that indeed the chromosome number was reduced by half in meiosis, and a new century would have arrived before the characteristics of individual chromosomes could be observed.)

Mendel's second law, known as the principle of independent assortment, goes on to state that the members of each pair are distributed independently when this segregation occurs. (Although this happened to be true for the chosen factors of Mendel's peas, we now know that factors don't necessarily operate so independently of one another. When crossing over occurs, those lying in close proximity on the same chromosome are more likely to be inherited together — to exhibit linkage — than those more widely separated, because break points are less likely to occur between them.)

Mendel gave a great deal of thought to the problem of the best plants on which to base his experiments. In describing his criteria for making that choice, Mendel wrote that it was most important that his plants "possess constant differing traits" and "easily lend themselves to protection from the influence of all foreign pollen during the flowering period." He chose peas, noting that

> Interference by foreign pollen cannot easily occur, since the fertilizing organs are closely surrounded by the keel, and the anthers burst within the bud: thus the stigma is covered with pollen even before the flower opens. ... Artificial fertilization is somewhat cumbersome, but it nearly always succeeds. For this purpose the not yet fully developed bud is opened, the keel is removed, and each stamen is carefully extracted with forceps, after which the stigma can be dusted at once with foreign pollen.

(Here Mendel is following in the footsteps of earlier religious practitioners, the priests of Assurnasirpal II, who donned their bird masks before artificially pollinating the date palms of Assyria, about 850 B.C.)

In 1854 Mendel went to his local seed dealers and bought thirty-four different varieties of common garden peas, mostly *Pisum sativum*, with which to start his test plots. He wanted to ensure that these plants were stable and would breed true for his chosen characteristics — that they would continue to replicate themselves faithfully through many future generations. After two years of testing, he settled on twenty-two varieties, which he planted religiously for eight years, meticulously making crosses with his forceps and carefully noting the results.

One of the first things he observed had been reported earlier by Gärtner in his massive book of 1849 entitled *Versuche und Beobachtungen über die Bastarderzeugung im Pflanzenreich*. This was that "it is entirely immaterial whether the dominating trait belongs to the seed or pollen plant; the form of the hybrid is identical in both cases." (Such felicitous reciprocity was not to manifest itself with the calicos, as has already been shown by the Punnett Squares presented earlier. Unhappy cat researchers were to learn this, to their confusion and dismay, some fifty years after Mendel, who had such a happy time with his peas.)

In the description of his experimental plan, Mendel is wonderfully clear about what he's up to:

It was the purpose of the experiment to observe these changes [in the hybrid progeny] for each pair of differing traits and to deduce the law according to which they appear in successive generations. Thus the study breaks up into just as many separate experiments as there are constantly differing traits in the experimental plants.

For his constantly differing traits, Mendel chose seven features covering various parts of the plant: the shape and color of the seeds, the color of their coats, the shape and color of the pods, the position of the flowers, and the length of the stems. Each feature could be manifested in one of two ways: Thus seeds were either round or wrinkled, yellow or green; the seedcoats gray or white; the pods inflated or pinched, yellow or green; the flowers axial or terminal; the stems long or short. Working with hybrids of these traits, Mendel got the following well-known results:

When plants which invariably produced wrinkled seeds were crossed with plants which invariably produced round seeds, all the plants of the ensuing generation produced round seeds only. The recessive trait of wrinkledness seemed to have been obliterated by the dominant roundness.

However, when Mendel planted these nice round seeds and allowed the resulting plants to fertilize themselves in their normal incestuous fashion, 25% of the resulting seeds turned out to be wrinkled.

There they were, lying side by side in the pod along with the round seeds, indicating that the wrinkled trait was still alive and well — and capable of reappearing in some mysterious fashion. And what's more, this constant ratio of 3 to 1 — three-quarters dominant vs. one-quarter recessive — reappeared repeatedly in Mendel's hybridization experiments with the rest of the seven traits as well.

Of course, Mendel wasn't content to stop there. He wanted to see what happened when he crossed plants that differed in two traits, or even three, rather than only one. So he chose plants with round yellow seeds (both traits dominant) and crossed them with those bearing wrinkled green seeds (both traits recessive) — and lo and behold, just as he predicted, all the peas of the next generation were round and yellow.

His next prediction was just as good. He believed that when these resulting plants were allowed to self-pollinate, they would produce the four different kinds of peas that were possible given the circumstances: round yellow, wrinkled yellow, round green, and wrinkled green. He also predicted the ratios in which these would occur — 9:3:3:1 — the doubly dominant ones, of course, being the most prevalent. And sure enough, it worked out just the way he expected.

His hypothesis that traits were represented by invisible factors that would act independently of one another was borne out beautifully by his experiments. The particular traits Mendel had chosen were probably the most pronounced and easily recognizable differences in the pea plants, and he had perhaps chosen them after observing their behavior for years. But unknown to him, they had a very important feature: *Pisum* has seven pairs of chromosomes, and he had selected one pair of factors from each. Thus in Mendel's many hybrids, his factors did indeed act independently of one another, since the different chromosomes on which they were represented never had an opportunity to cross over, exhibit linkage, and mess up his nice, neat results.

(The geneticist/statistician R.A. Fisher, reexamining Mendel's data in 1936, points out that his results are so neat as to be questionable. He suggests, with deferential care, that Mendel "knew very surely what to expect, and designed [his experiments] as a demonstration for others rather than for his own enlightenment." He also reminds us that it's sometimes hard to tell the round from the wrinkled and says that "it must be concluded that he made occasional subconscious errors in favor of expectation." Fisher concludes, however, by ruling out any deliberate effort at falsification, and the slight cloud that he cast in no way diminishes the beauty and importance of Mendel's work.)

Being a good scientist, Mendel wanted to know if his hypothesis was generally true — whether, for example, it would work for beans as well as for peas. So he did two small experiments with *Phaseolus*, a scarlet-runner-like bean plant, and the first "gave fully concordant results." The second, however, "was only partly successful" because of

> a remarkable color change in the blossoms and seeds of hybrids. ... Besides the fact that a union of white and crimson coloration produces a whole range of colors from purple to pale violet and white, it is also striking that out of thirty-one flowering plants only one received the recessive trait of white coloration, while in *Pisum* this is true of every fourth plant on the average.

Others had noticed this range of color in hybrids and had used it to support the current theories of inheritance, which ran toward blending of fluids, as of two different-colored liquids poured together. Mendel, however, persists with his independent-factor theory, even in the face of these disturbing results. He proposes that

> these puzzling phenomena, too, could probably be explained by the law valid for *Pisum* if one might assume that in *Phaseolus multiflorus* the color of flowers and seeds is composed of two or more totally independent colors that behave individually exactly like any other constant trait in the plant.

He then goes on to explain mathematically how if each color were represented by two pairs of factors, different combinations could be produced in the hybrids, and these could then generate the gradation of color that he observed; further, the colors would be distributed in unequal proportions.

Mendel may not have had an accurate understanding of *Phaseolus*, but in holding out against blending and sticking to his beliefs about discrete factors, he proposed yet another important new idea: that there may not be a one-to-one correspondence between traits and factors. Rather, many factors may be working in conjunction with one another to produce a singular effect — like flowers of many hues or George's wonderfully complicated three-colored coat.

Darwin trips over the calicos

With all my reading into the history of George's ancestors, I had failed to discover any clue to when people first noticed that almost all calicos are female. It must have been within recorded history, but it's apparently a piece of history that no one bothered to record. In fact, surprisingly little has been written about calicos in general, either now or at any time in the past.

It occurred to me, though, that Darwin probably had some knowledge and opinions about this matter and that he might even refer to theories earlier than his own. I rescued my ancient, dusty copy of *The Descent of Man* from its exile on the highest shelf of my home library and began to read it for the first time. Looking in the back, I found that this 1874 volume was from Burt's Home Library, a series that advertised itself as "Popular Literature for the Masses" (bound in cloth, with gilt tops, for $1.25 apiece). It didn't look much like Louis L'Amour to me.

Looking in the front of this second edition, I began, of course, to read the Preface. There Darwin speaks briefly of "the fiery ordeal through which the book has passed" and ends with the note that "it is probable, or almost certain, that several of my conclusions will hereafter be found erroneous; this can hardly fail to be the case in the first treatment of a subject." Fortunately, an extremely detailed index was provided, and it didn't take me long to discover that one of his erroneous conclusions had to do with tortoiseshell cats. Darwin, like others before him, was puzzled about why "it is the females alone in cats which are tortoiseshell, the corresponding color in the males being rusty-red."

Mivart, the famous biologist, has similar thoughts in his strange book *The Cat*, published in 1881:

> It appears that the sandy tom cat is the male of the breed of which the tortoiseshell is the female ... . This fact is very interesting, because the sexes of cat-like animals are similarly coloured.

He goes on to say that

Sometimes, however, sandy cats are female, and there is at least one good instance of a true tortoiseshell tom cat. Such cats, indeed have not unfrequently been offered, by letter, to the Secretary of the Zoological Society, at very extravagant prices.

Reading this, I wanted very much to see the tone, style, and content of such letters and to know what was meant at that time by "extravagant prices." So I dashed off a note to the current Secretary, asking whether copies of such letters could be made available. As I addressed an envelope to Regent's Park, I was reflecting on the long continuous history of this organization — the Zoological Society of London was founded in 1826 — and the important repository of knowledge that it consequently represents. This image was shattered, however, by the following short reply:

Unfortunately it is not possible to help you with regard to letters to the Society offering tortoiseshell cats for sale. The Society's records were destroyed during the last war and no such letters have survived.

As I grieved slightly over my lost cat letters, I wondered how many similar responses various Secretaries had been obliged to write and in answer to what sorts of (no doubt much more important) questions. It was a sad reminder of the periodic, catastrophic discontinuities of history that the violence of our species — and sometimes of nature — produces.

Even without the letters, however, it was clear to me that the Victorians were largely unaware of the existence of any Georges but did know and puzzle about the fact that calicos were usually female. As to how they explained this strange phenomenon, Mivart provides no answers, but Darwin worries out loud about this and other sex-based inequities in a section entitled "Inheritance as Limited by Sex." He considers "whether a character at first developed in both sexes could through selection be limited in its development to one sex alone." Finally, he proposes that the two following rules seem often to hold good — that variations which first appear in either sex at a late period of life tend to be developed in the same sex alone; while variations which first appear early in life in either sex tend to be developed in both sexes. I am, however, far from supposing that this is the sole determining cause.

As examples to support his rules, Darwin reminds us that

in the various domestic breeds of sheep, goats, and cattle the males differ from their respective females in the shape or development of their horns, forehead, mane, dewlap, tail and hump on the shoulders; and these peculiarities, in accordance with our rule, are not fully developed until a rather late period of life.

But as an admission that his rules are neither complete nor foolproof, he goes on to say that "the tortoiseshell color, which is confined to female cats, is quite distinct at birth, and this case violates the rule."

As genetics started to emerge as a science at the beginning of the twentieth century, the calico cats caused the violation of a lot of other rules as well.

A pair of giants

It's really too bad that Darwin and Mendel never met or even corresponded; it would certainly be interesting to learn what these two giants of the nineteenth century thought of one another. Although Mendel was a peasant and Darwin a member of the aristocracy, they had many things in common: Both suffered from poor health, enjoyed seclusion, and were passionately interested in the details of living things. Both thought deeply and carefully and embarked on studies that were to last for years and years. As it is, they had a very one-sided relationship: Mendel was quite interested in Darwin, but Darwin (as far as anyone knows) was totally unaware of Mendel's existence. This is puzzling because Darwin seems to have read everyone, even the ancients, but somehow Mendel escaped his omnivorous attention.

Mendel mentions Darwin's work in his letters to Carl von Nägeli and points out errors of various sorts. Darwin was of the opinion that a single grain of pollen was insufficient to fertilize an egg, but Mendel had tried it and knew better. Mendel had read Darwin's *The Variation of Animals and Plants Under Domestication* and says that some of the views concerning hybrids "need to be corrected in many respects."

Mendel had also read, and even underlined, *The Origin of Species*. We know this because his German copy, published in 1863, is on display at the Moravian Museum in Brünn. But we'll never know what he thought about many Darwinian ideas, because he never wrote about them — he was clearly not in a position to get mixed up in all that anti-religious controversy the *Origin* inspired. Nevertheless, in his introductory remarks describing his lengthy experiments with peas, Mendel says,

> It requires a good deal of courage indeed to undertake such a far-reaching task; however, this seems to be the one correct way of finally reaching the solution to a question whose significance for the evolutionary history of organic forms must not be underestimated.

Courage indeed. Here's a nineteenth-century monk speaking in terms of "the evolutionary history of organic forms." The *Origin* had caused a furor when it first appeared in 1859, but clearly Mendel had been thinking in these terms even earlier, having laid the groundwork for his experiments before 1854.

(The first edition of *The Origin of Species* contains no mention of the Creator, but later Darwin relents, perhaps under pressure from his very religious wife, and allows Him to creep into the closing sentence of later editions. Darwin fails to award Him all the animals of Noah's ark but does allow Him one or two forms with which to start the ball rolling. He seems to be echoing Linnaeus's 1764 references to simplicia and composita in this surprisingly poetic ending: "There is grandeur in this view of life, with its several powers, having been originally breathed by the Creator into a few forms or into one; and that, whilst this planet has gone cycling on according to the fixed law of gravity, from so simple a beginning endless forms most beautiful and most wonderful have been, and are being evolved.")

Darwin also tells us, in a footnote, that Aristotle had beaten them all to the punch in his *Physicae Auscultationes* by pointing out that the rain does not fall in order to make the corn grow, any more than it falls to spoil the farmer's corn when threshed out of doors. He then proceeds to extend this exposition about lack of purpose to the different parts of the body:

> So what hinders the different parts from having this merely accidental relation in nature? ... Wheresoever, therefore, all things together (that is all the parts of one whole) happened like as if they were made for the sake of something, these were preserved, having been appropriately constituted by an internal spontaneity; and whatsoever things were not thus constituted, perished, and still perish.

Natural selection in a nutshell (even in this extremely awkward translation by Darwin's friend Clair Grece).

But many others of Darwin's generation had harsh words for natural selection, and Mivart was one of them. He writes in his Cat book,

> The notion that the origin of species is due to "Natural Selection" is a crude and inadequate conception which has been welcomed by many persons on account of its apparent simplicity, and has been eagerly accepted by others on account of its supposed fatal effects on a belief in Divine creation.

Mendel thought that traits were inherited through independent factors that were passed on from generation to generation through strict mathematical laws. Darwin had no image of Mendel's factors but, in a revival of Aristotle's pangenesis, believed instead that "gemmules" circulated in the blood throughout the body. Darwin thought these mysterious and invisible entities were feature bearers that not only made it possible for long noses to appear in many generations of a family but also allowed for the inheritance of acquired characteristics, like housemaid's knee. (In support of the gemmule theory, Darwin's cousin Francis Galton transfused hundreds of rabbits but never produced the desired results. Mivart also believed in the inheritance of acquired characteristics; he cites the case of a female cat who produced kittens with stumps for tails only after her own tail was cut off near the root when she was run over by a cart.)

One of the most obvious differences between Mendel and Darwin was in their style of writing. Where Mendel is crisp and concise, Darwin is obscure and tedious. He drones on and on, from one difficult and convoluted sentence to another, presenting speculation and anecdotal observation, slowly building a lengthy argument and a mountain of facts with which to shake the foundations of Victorian thought.

Darwin was aware that writing was not his thing and was extremely surprised at the success of his books. (He had been known to grumble that if a bad arrangement of a sentence was possible, he would be sure to adopt it, and he had said, with surprise, while rereading the *Origin*, "it is a very good book, but oh! my gracious, it is tough reading.") He opens the Introduction to *The Descent of Man* as follows:

> During many years I collected notes on the origin or descent of man, without any intention of publishing on the subject, but rather with the determination not to publish, as I thought that I should thus only add to the prejudices against my views.

Mendel, by contrast, introduced terminology and notation that are still in use today. He chose the German equivalents of *dominant* and *recessive*; he also invented the notation of using upper- and lower-case versions of the same letter (Aa) to represent the dominant and recessive versions of the same trait.

Because of his mathematical training at the University of Vienna, Mendel's paper is filled with what appear to be mathematical expressions and even a few equations. Most, like

$$\frac{A}{A} + \frac{A}{a} + \frac{a}{A} + \frac{a}{a} = A + 2Aa + a$$

make no sense when taken mathematically, and this may have accounted in part for the lack of comprehension that Mendel encountered. But here, as he carefully explains, the top part of the fraction represents a trait in the pollen cell, the bottom part a trait in the seed cell. Mendel is trying to tell us, as Gärtner did, that $\frac{A}{a}$ is the same as $\frac{a}{A}$; it doesn't matter which part made the contribution to the hybrid, the result is the same.

He's also succinctly expressing the idea that plants that receive two dominant factors $\frac{A}{A}$ will exhibit this dominant trait (A) and plants that receive two recessive factors $\frac{a}{a}$ will exhibit this recessive trait (a). And since those that have both a dominant and a recessive factor (the two that are Aa) will exhibit the dominant trait (A) as well, the righthand side of the equation follows his famous 3:1 ratio.

Mendel's most important contribution may have been to apply, for the first time, serious statistical measures to biological entities. Ironically, his clear but novel notation may have rendered his results incomprehensible. And so while everyone was reading Darwin's voluminous prose, and getting very hot under the collar about his ideas, Mendel's short paper was lost and forgotten. Mendel was unknown, of course, and had no hope of reaching Darwin's masses, but it's particularly sad that his beautiful paper, with its precision and clarity, failed to communicate its important messages even to the other scientists of his time.

Cats are not peas

No one will be surprised to hear that cats have a different genetic make-up from peas. Cats have thousands and thousands of genes, most of which are busy controlling the internal functions that make the cat go (and guaranteeing that it's a cat that gets built, not a pea plant or a tree). A few of these genes, however, have highly visible effects. They're the ones that are in charge of what the cat looks like; they control, among other things, the color and type of the hair and the general structure of the body. It's these genes that enable us to understand some things about the underlying genetic structure merely by observing external variations on the theme.

Theoretically, following Mendel's law of independent assortment, all coat colors, coat types, and body types can occur in any combination. Even eye color, which ranges from a coppery orange, through yellow, hazel, and green, into several shades of blue, is remarkably independent of coat color. Still, the Siamese cats usually have deep blue eyes, and pure white cats usually have light blue or orange eyes (sometimes one of each), so it's clear that coat color and eye color are linked to some extent. Certain other linkages between genes, especially those that are close together on the same chromosome, will commonly occur, but in general the possibilities for variation seem endless — there's no knowing what combinations may yet arise.

The Book of the Cat devotes several pages to tortoiseshell and calico coats and displays twenty-four different arrangements of colors and tabby types that might co-occur. Alongside each beautifully drawn simulated pelt, the relevant set of genes is helpfully provided. I looked through the examples carefully but soon realized that George's type is missing — not because he's a male calico (which is yet another problem), and probably not because his type is particularly rare, but just because not all combinations can be listed.

Some major cat genes do their work in a simple, straightforward fashion, like those for peas that decide whether the seed will be round or wrinkled, yellow or green. Such a simple pair controls hair length: The basic wild-type gene calls for short hair, whereas the recessive mutant that arose in Russia calls for long hair. Following Mendel's convention (but adding modern italics), short hair is given the symbol L (upper case to indicate that it's dominant), and long hair is given the symbol l (matching lower case to indicate recessive). You can tell just by looking at Max's long, silky hair that he must have two of the latter (ll); if he had only one, the contrasting L from the matching chromosome would dominate and he would be a short-haired cat like George. (There's no way to know, though, whether George is LL or Ll, as short hair will result in either case.)

A similarly simple case can be constructed with regard to the agouti gene *A* and its non-agouti counterpart *a*. The dominant *A* causes banded agouti-colored hair to be maintained between the tabby stripes; the recessive *a* specifies non-banded hair the same color as the tabby stripes, thus producing a uniform-colored coat. Since Max is solid black and shows no agouti banding, he must be *aa*. But George's agouti bands indicate that he may be either *AA* or *Aa*; we can't tell which by looking at his coat since the dominant *A* always wins. It works just like the round and wrinkled peas.

But things aren't always so neat, as Mendel found out for *Phaseolus*. To account for the unexpected range of colors that appeared when plants with white and crimson flowers were interbred, Mendel had the wisdom to postulate that two or more totally independent factors might be contributing their two cents worth to the final hue. What he didn't realize was that the voting might be unfair — that some genes could actually mask the effects of others, even those on different chromosomes. Relationships between genes are actually much more deeply and richly interconnected (more "intertwingled," to borrow some wonderfully descriptive jargon from computer science) than Mendel ever imagined.

An extreme example of such intertwingling in cats is provided by the dominant white gene *W*. This gene is so powerful that even one of them is enough to mask all other colors. Thus you can't find out very much about the color genes of a pure white cat by simply inspecting it. The cat may be *WW*, it may be *Ww*, or it may have albino genes of some sort; there's no way to know, except by extensive breeding, exactly what has caused its whiteness. Even worse, there's no way to know what other color genes may be lurking at other locations, perhaps on other chromosomes, since they're all masked completely by the effects of *W*. Thus breeders may have all sorts of surprises in store for them when dealing with totally white-haired cats.

Let's take a second look at Max's tabby genes and what they're up to. Of course, as some of you may have noticed, the tale we've told didn't really cover the waterfront. Maybe his dark stripes wouldn't show against all that black fur. But Max is white where he isn't black, so why don't the tabby stripes show all over his white tuxedo front? It clearly isn't *W* that's at work here — it can't be or Max would be white all over. This is the handiwork of *S*, another dominant mutant gene, which calls for white spotting. As with *W*, the areas that lie under the control of *S* are guaranteed to have solid white hairs, no matter what other color genes may be present — thus no tabby stripes can appear. (George must also have an *S* because his white areas are pristine like Max's.)

The white-spotting gene *S* also manifests "variable expression." This means that the amount of white acreage produced depends on whether one or two of these genes are present. Since Max and George are less than one-third white, it's likely that each has only one *S* instead of two; thus they're both probably *Ss*, with the wild-type gene *s* voting for no white spotting at all.

In pure tortoiseshell cats, which have no areas of white and hence no *S*, the black and orange hairs are sometimes intermixed so closely that they produce a motley, mosaic appearance. In calicos, which have various amounts of white and hence at least one *S*, another surprising action of the dominant white-spotting gene is to interact with the genes at the orange locus to produce large, sometimes widely separated, patches of orange (*O*) and non-orange (*o*); with non-orange manifesting itself as black. In Japanese bobtails, which are mostly white and hence *SS*, these dense orange or black patches may be so sparse and driven so far apart that only one of the two colors is represented. The gene for the absent color is still present, however, harboring another surprise for unwary breeders.

It would have been a big surprise for Mendel as well. He was great, but he probably wouldn't have been able to figure out all this intertwingled stuff, even if he hadn't been made abbot.

The rediscovery

So what did happen to Mendel's famous paper about his round and wrinkled peas? It's certainly tempting to cast the description in roman-tic terms: an exhausted abbot; a remote monastery; a moldy piece of parchment, rolled tightly, tied with a ribbon, and stashed away in a secret compartment. One entertains Umberto Eco–like images of quills, high stools, long tables, bent backs, and intricately illuminated manuscripts.

Of course it wasn't like that at all. Although Mendel certainly wrote by hand (and may have used a quill for all I know), by 1865, when his forty-eight-page pea paper was ready to be published, the printing press had already been cranking out copies of things for over four hundred years. And these copies weren't usually hidden, they were distributed: The Proceedings of even such a small and obscure group as the Brünn Natural History Society were circulated to more than a hundred and twenty libraries throughout Europe, and eleven copies of Mendel's pea paper reached the United States before 1900.

Mendel himself was given forty reprints to distribute. Two of them went to the leading botanical professors of his time: C. von Nägeli at Munich and A. Kerner von Marilaun at Innsbrück. Nägeli totally failed to appreciate the pea paper — he speaks of it "with mistrustful caution" despite the seven-year correspondence with Mendel that it initiated. (Later, however, he showed considerable interest in Mendel's *Hieracium* investigations, to which he contributed advice along with numerous plants.) Kerner totally missed his chance by never even bothering to open Mendel's offering; his copy was found after his death in 1878 with its pages still uncut. No one knows the names of the other lucky recipients.

Eventually there were to be a few citations as well. In 1881 a long, well-organized bibliography of work on plant hybrids called *Die Pflanzen-Mischlinge* was published in Germany. It mentions Mendel fifteen times and contains pointers to his papers under the headings *Pisum* and *Hieracium*, but it devotes very little text to his discoveries. Under *Pisum* it reports that "Mendel thought that he found constant ratios between the hybrid types"; under *Hieracium* it simply says, "The hybrids are polymorphic, according to Mendel's experiences, but the individual forms usually produce true-breeding seeds." The ninth edition of the

Encyclopaedia Britannica, which appeared between 1881 and 1895, mentions Mendel briefly in its article on hybridism, and his paper was listed in the *Royal Society Catalogue of Scientific Papers*.

Not much publicity to be sure, but enough of a trace for serious scientists who know the value of a thorough literature search before publication. Three such men at the turn of the century were Carl Correns, Erich von Tschermak-Seysenegg, and Hugo de Vries. All had undertaken hybridization experiments with no knowledge of Mendel's or each other's work, and they had come to similar Mendelian conclusions almost simultaneously. Independent assortment was clearly an idea whose time had come. All three of them published papers in 1900, giving due reference and reverence to the long-forgotten Moravian monk.

Correns and Tschermak were both Germans who had worked with *Pisum*; they found their leads to Mendel's papers in 1899 through *Die Pflanzen-Mischlinge*. (Correns had also heard about Mendel personally from his teacher Nägeli, but only with regard to his work on *Hieracium*.) There are three different versions of how de Vries, a Dutchman, first encountered Mendel. Two slightly conflicting tales are told by de Vries himself; a totally different and much more interesting one is told by his student Stomps ten years after de Vries's death.

De Vries himself, when asked in 1924 to contribute to the historical record of the rediscovery, said that he had found a reference to Mendel in the literature list at the back of an 1895 book by Bailey entitled *Plant Breeding*. Unfortunately, this first edition has no bibliography. To complicate matters further, in its fourth edition, Bailey quotes a 1908 letter of de Vries in which he thanks Bailey for his 1892 article on "Cross-Breeding and Hybridization" and says he found Mendel's paper in *its* bibliography.

These and other confusions cast a slight cloud over de Vries: There were suggestions that he had read Mendel at an earlier date than he admitted and that in his first publications he had even attempted to suppress Mendel's name. Indignant at this tarnishing of his teacher's reputation, Stomps wrote a short article in the *Journal of Heredity* in

1954, giving the following account of the rediscovery as told to him personally by de Vries:

> In 1900, at just the time he was about to publish the results of his experiments he received a letter from his friend Professor Beyerinck at Delft, reading thus:
>
> > "I know that you are studying hybrids, so perhaps the enclosed reprint of the year 1865 by a certain Mendel which I happen to possess, is still of some interest to you."
>
> Dr Vries read the paper and found that the results of his experiments, which he had believed to be quite new, had already been reported 35 years before.

Of course one now wants to know about Beyerinck. Had he read the paper? Did he understand it? Had Mendel sent it to him personally? Was he just cleaning out his office one rainy afternoon, dispensing various extraneous documents in appropriate directions? Did he have any idea what a bombshell he was dropping on de Vries? (I don't know the answer to any of these questions, but I do know that Beyerinck later lamented that he would have been the first to rediscover Mendel, five years before de Vries, if only he hadn't abandoned his hybridization experiments in Wageningen to become a bacteriologist at the Dutch Distillery Works in Delft. It probably seemed like a good career choice at the time.)

Stomps cites as proof of this remarkable tale the interesting fact that

> After the death of Beyerinck ... his family sent the reprint in question again to our institute, this time to me as its director, with the words that the right place would be the library of the Botanical Institute at Amsterdam, where indeed one can see it today in a special showcase.

A fitting end to a long and lonely odyssey. It would be a nice pilgrimage to go and look at Mendel's famous paper in such a suitable setting; it would also be nice to know the itineraries of the remaining thirty-seven reprints.

6

What Did They See
and When Did They See It?

The botanists beg to differ

THE NEW SCIENCE OF GENETICS WAS BORN FROM THE EXPLOSION OF IDEAS that occurred with the rediscovery of Mendel. But it wasn't all smooth sailing. Ironically, it was the botanists who were the most hostile to, and held out the longest against, the new theories of "Mendelism." Even that august British journal *Nature*, in which the announcements of many famous discoveries have first appeared, refused for several years to publish papers on the subject, choosing instead to support the rival theories of the Biometricians, who constituted the opposition. To understand what happened at the beginning of the twentieth century we must dip back briefly into the nineteenth, to find out who saw what when and to understand the prevailing beliefs that these sights engendered.

In 1828, shortly after the birth of Mendel, the English botanist Robert Brown (of Brownian motion fame) saw that cells have molecules moving around inside them; in 1833 he also discovered that the cell has a nucleus.

By 1842, about ten years before Mendel entered the University of Vienna, Nägeli actually saw mitosis in action. He watched as a cell split in half to produce two new ones, and he realized that the nucleus itself also split to form two new nuclei. During the process Nägeli caught a vague glimpse of the chromosomes, which he called the German equivalent of transitory cytoblasts. Neither he nor anyone else at that time had any useful theories about what the chromosomes were good for. Historians disagree (no big surprise) about Mendel and chromosomes — some say he had no knowledge of them. But if Nägeli had seen and named them, it's hard to believe that Mendel didn't know about them.

By 1873, when Mendel had already done five years of his sixteen-year abbot sentence (and probably wasn't paying too much attention to anything else), a German biologist named Schneider, working with flatworms, saw the lining-up phase of mitosis as the chromosomes gathered on the equatorial plate; then he watched the poling-up phase take place as they migrated to the opposite ends of the cell.

In 1883, a year before Mendel's death, the initial doubling-up phase of mitosis was observed by Walther Flemming, who was studying the larvae of salamanders at the time. He was the one who first saw that the transitory cytoblasts split longitudinally to replicate themselves. Fortunately, he gave the resulting thread-like structures the more manageable name *chromatin*, which was soon modified to *chromosome*. (This term, which means "colored body," describes what they look like when stained for microscopic viewing.)

A few years later, Flemming became one of the first to view the second division of meiosis, in which the number of chromosomes is halved. Flemming reported that although most salamander cells contained 24 chromosomes, their egg cells contained only 12, but he failed to recognize the significance of this important fact. If Mendel had still been around to hear of Flemming's discovery, he surely would have realized that it confirmed his principle of segregation, postulated more than thirty years earlier.

Mendel being dead, the task of noticing and heralding the meaning of the reduction division fell to a fiery evolutionist named August Weismann. Independently of Mendel, of whom he had no knowledge, Weismann proposed the theory that the chromosomes were composed of a large number of discrete and different "ids," which were the bearers of hereditary traits. Whereas others thought that the chromosomes were all identical — being barely visible, they all looked alike — Weismann was convinced that they were all different and that they segregated into two groups, each of which formed a new nucleus. By Weismann's method, all the gametes would be different from one another and could pass along a wide variety of trait-bearing ids to the next generation. He was the first to call the meiotic process a reduction division and described its purpose as "the attempt to bring about as ultimate a mixture as possible of the hereditary units of both father and mother."

Given the different preconceptions and biases researchers brought with them to their still inadequate microscopes, it's not surprising that their eyes sent different messages to their brains about the tiny objects they were straining to see. During the 1890s, most biologists (many working with sea creatures) didn't agree with Weismann and his students (who were working mostly with insects). They could see that the chromosome count was indeed reduced by half, but they believed the reduction was merely quantitative, not qualitative, since all the chromosomes were identical. Thus they thought that all the gametes of an organism would be identical as well. On these grounds they claimed that "a reduction division in the sense of Weismann does not occur." The botanists were the most adamant in this disagreement. Strasburger, a famous plant person, wrote in 1894, "There is no reduction division in the plant kingdom nor anywhere else." And that seemed to settle that.

Since this was the prevailing attitude at the turn of the century, it's amazing that anyone was able to notice the connection between Weismann and Mendel when Mendel's pea paper was finally rediscovered in 1900. Even among the three discoverers (all botanists), only Carl Correns really understood the importance of Mendel's contributions. He saw right away how round and wrinkled, gold and green, fit neatly into a qualitative rather than a quantitative reduction division, and he realized that hereditary traits must indeed be resident in the chromosomes, just as Weismann had postulated fifteen years earlier.

In England, one of Mendel's strongest proponents was the English zoologist William Bateson. In a delightful short memoir of these early days, R.C. Punnett (of Punnett Square fame) describes how he and other enthusiastic young disciples helped Bateson do his genetic experiments around the house. First in an upstairs bedroom, and later along the garden paths, they raised and crossed various kinds of poultry in portable brooders (which occasionally caught fire). They also grew and crossed thousands of sweet peas, having to move their crops to a nearby field when Bateson's wife protested that the yard space was needed for a vegetable garden. Results were duly recorded by Mrs. Bateson in lab books, which also included notes of bets made by the researchers about expected outcomes. (One of those betting was Doncaster, who was later to expend lots of effort on the problem of the male calico cat.)

Bateson was the first in his country to announce the rediscovery, which he learned about on a train taking him to a meeting of the Royal Horticultural Society. Understanding its implications immediately, he revised his lecture *en route* in order to bring Mendel to the attention of the assembled multitudes. Bateson later translated the pea paper into English and became for the dead Mendel what Thomas Huxley had been for the live but reclusive Darwin, a noisy and successful apologist.

Mendel's principles were starting to emerge from oblivion, no matter what some of his fellow botanists had to say about it.

The great lubber grasshopper

In America, Thomas Montgomery theorized that matching chromosomes paired up during the reduction division, and the famous biologist Edmund Wilson set his student Walter Sutton loose to test this theory on *Brachystola magna*, the great "lubber grasshopper." *Brachystola's* chromosomes come in a wide variety of sizes and shapes, so it was possible, even at the turn of the century, to see both differences and similarities among them. By 1902 Sutton was able to confirm that the matching maternal and paternal chromosomes do indeed pair up during meiosis and then are isolated in the production of the sex cells. As Wilson said, "this gives a physical basis for the association of dominant and recessive characters in the cross-bred ... exactly such as the Mendelian principle requires."

McClung, another student of Wilson, also worked with the great lubber grasshopper and reported that it produced two different types of sperm cells: one with 11 chromosomes and one with 12. (He turned out to be right: *Brachystola* is one of the few organisms, like the creeping vole, which has an uneven number of chromosomes: one more in the female than in the male.) He termed the extra one the accessory chromosome and theorized that it was responsible for sex determination. He argued quite reasonably that if organisms were to be divided into two camps, formed by two different sorts of sperm, then sex was the only sensible dividing line between them.

McClung's famous paper entitled "The Accessory Chromosome — Sex Determinant?" (also published in 1902) provides an extensive historical survey in which he points out that Hermann Henking, working in Germany with male *Pyrrhocoris* bugs, had described this accessory chromosome way back in 1891. Since Henking didn't think it was actually a chromosome and didn't know what it was, he called it Doppelelement X (for unknown), and that's how the X-chromosome got its boring name. When its tiny counterpart was finally seen in organisms other than *Brachystola*, there seemed to be no choice but to name it Y, which Edmund Wilson did in 1909.

In those days Wilson was still proclaiming that "external conditions" were in charge of sex determination and that it was "certain that sex as such is not inherited." Despite his mentor's pronouncements, McClung believed that the function of the accessory chromosome was to provide the extra oomph necessary to change an ovary into a testis. As it turned out McClung had it backwards: In grasshoppers as well as fruit flies, it's the absence of an X, rather than its presence, that produces maleness. Other turn-of-the-century confusions arose because of the wide variety of organisms that were being studied — besides grasshoppers and fruit flies, there were bees and wasps, spiders, butterflies, sea urchins, salamanders, birds, cats, and people — with the assumption that the same sexual mechanisms were probably at work in all of them.

Sex and the single slipper shell snail

In all mammals and most insects, the females are XX and the males XY, the sex of the offspring being determined by the male. For birds, however, as well as for moths, many fishes, salamanders, frogs, newts, and snakes, it's the other way around: The females are XY and the males XX, the sex of the offspring being determined by the female. Different boring terminology, with an obvious etymology, is sometimes used for them: birds in particular are often described as having a ZZ/ZW system in which the males are ZZ, the females are ZW, and the W is assigned the job of sex determination.

Like its counterpart the X, the Z is a sturdy chromosome heavily populated with genes; in the case of birds, many are dedicated to providing brilliant plumage for the males and perhaps song sequences as well. Like its counterpart the Y, the W has shrunk to insignificance because it has little to do but specify sex.

Some creatures — the ameba, for example — are sexless and reproduce simply by splitting into two identical parts through mitosis. Those creatures who do enjoy the benefits of sexual reproduction accomplish it in a bewildering myriad of ways. For some species of insects, fishes, and lizards, the population consists of females only; some are hermaphrodites with working reproductive organs of both sexes; some fish change their sex throughout their lifetime; for others, the temperature of the environment is a controlling factor in sex determination.

One of the most curious schemes of all is that of the slipper shell snail, which is initially male only. But its sex changes, depending on where it happens to land as it sinks to the bottom of the sea. Young males become female when they hit bottom, unless they happen to land on a female, in which case they remain male — but if they become detached from the female for some reason, then they become female. Thus their sex keeps changing, depending on the environment, in some complicated scheme to increase the probability of mating. Some worms, in which the males are tiny parasites of the females, have a similar system: Those worm larvae that happen to become attached to the proboscis of an adult female become male; all others sink to the bottom of the pond and become female.

The budding cytologists at the beginning of the twentieth century, of course, hadn't discovered all these things yet and were pretty much in a muddle where sex determination was concerned. However, they all knew a lot about mitosis and meiosis, believed that the chromosomes were the transmitters of hereditary information, saw that they came in maternal and paternal pairs, agreed that sex cells were likely to be different from one another, and thought that sex was probably one of many traits encoded somehow in the chromosomes.

Fecund little creatures

Probably everyone knows about *Drosophila*, the fruit fly that came to be the work horse — to mix genera — of the early geneticists. In fact, when you read *The History of Genetics* as told by that famous pioneer of the field A. H. Sturtevant, you come away with the impression that "in the beginning was *Drosophila*" and nothing else (except Mendel's wonderful peas) mattered very much. Sturtevant provides an interesting footnote about mice having their tails cut off for twenty generations in order to test the supposed inheritance of acquired characteristics (it didn't work), but nowhere is there even a mention of the cats who were also busy giving their all so that people's curiosity about how heredity worked might be satisfied.

Fruit flies are extremely simple to obtain, and they propagate like mad. Since they lay hundreds of eggs at a time and a new generation makes its appearance every two weeks, it's not surprising that by 1916 Morgan and Sturtevant had happily bred over half a million of the little pests. (*Drosophila*, which means "dew lover," was probably what Aristotle was identifying when he described a gnat produced by larvae engendered in the slime of vinegar.)

Their fertility, along with the large size and small number of their chromosomes (they had only 4 pairs), made them very desirable subjects. *Drosophila* didn't mind being confined in small spaces like half-pint milk bottles, their breeding could be carefully controlled, and no animal rights groups were likely to arise complaining about their treatment. Like Mendel's peas, they provided some nice contrasts that were easy to see (red eyes vs. white, for example), and soon over a hundred different factors were being analyzed.

The famous Fly Room at Columbia, in which *Drosophila* were bred and studied from about 1910 to 1927, measured only 16 by 23 feet. Besides all those flies, it contained the eight desks of the excited young geneticists who made the first chromosome maps, discovered sex-linked genes, and first proposed non-disjunction (here termed reluctance; see Figure 7). Many extolled *Drosophila* as the perfect medium for experimentation.

Nowhere in the *Drosophila* literature does one come upon laments regarding the difficulties of working with these willing and fecund little creatures. Sturtevant does mention that early results were poor in that they rarely came close to the expected Mendelian ratios of 3:1, but this was recognized as being due to different mortality rates in larval and pupal stages before the counts were made. He also gives an amusing description of a particularly unusual fly that was being examined by Mrs. Morgan when it recovered too soon from the anesthetic and flipped itself off the microscrope stage onto the floor. I imagine her scrambling about on all fours under all those desks, desperately searching for this special gnat. Failing, she reasoned that flies go toward the light when disturbed and was lucky enough to find her special specimen on the window, recognizing it because of its peculiarities.

Utterly bad mothers

Cats were a different story. Calicos in particular are always somewhat hard to find, calico males very rare, and the fertile males almost nonexistent. Some useful breeding experiments could be carried out even in the absence of calicos (such as crossing orange females with black males, and vice versa), but results were a comparatively long time in coming. Even if animals of the appropriate colors could be found, they couldn't necessarily be induced to mate; and when they did, they were likely to produce only one litter of a few kittens every year. They were no competition for *Drosophila*.

Some cats just didn't feel like breeding or mothering at all and caused some rather desperate writing to appear in the midst of serious scientific papers. The following remarks, published in the *Journal of Genetics* in 1924, concerned Siamese and white Persian cats being used in experiments designed to discover the genes controlling their eye color and hair type.

> They do not like being out of doors, they cannot be kept in pens, and will not thrive without the company of man. They need the cosiness and warmth of a human dwelling and must be treated as pet animals. The females are shy and mating is often difficult. ... The white Persians are perhaps still more difficult to breed than the Siamese. The weakness of these cats is very striking and their females are utterly bad mothers; they often eat their kittens at birth or starve them to death a few days after, being wearied of their nursing duties. ... Nearly all the members of this family [which was a cross between the difficult Siamese and the even more difficult Persians] showed somatic and mental defects (sterility, deafness, dirtiness, inability to withstand the simplest difficulties of a cat's life). This caused much trouble in the course of my experiments.

Others mated all too freely, especially those in the hands of the breeders from whom the early cat geneticists often obtained their data. A critique of some research from 1913 cautions:

> but his data were collected from breeders for the Cat Fancy. ... Thus there is not a very firm foundation of fact on which to erect such a weighty superstructure of hypothesis.

This situation allowed investigators first to malign and then to ignore any upsetting information from their rivals. Then as now, just determining the sex of a kitten wasn't easy (unless drastic measures were taken), and this fact, like the breeder's doubtful records, provided yet another excuse for discounting unwelcome data.

And so it went in the cat world. While the many drosophilaphiles were writing happily about their interesting new results, the small number of cat people were producing literature with a high incidence of words such as *regret, difficulties, untimely, doubtful, questionable,* and *unfortunately.* Thus my earlier image of geneticists surrounded by flies, hunched over their inadequate microscopes, desperately straining their eyes to see the sex chromosomes of *Drosophila* has been supplanted by an image of geneticists surrounded by fleas, tearing their hair, and cursing the behavior (or lack thereof) of *Felis domestica.*

7
The Early Calico Papers

Rock 'n' roll

IN MID-OCTOBER OF 1989, GEORGE IS ENGAGED IN HIS SEMI-ANNUAL vanishing act and has been gone for several days. Max is moping in the laundry room, sprawled with all four legs extended on top of the slick white surface of the washing machine, trying to stay cool in the unseasonable heat. We are upstairs, hunched together over a computer screen, attempting to dispel the worry that George's absences always cause by burying ourselves in some especially tedious work.

With a sudden roar the house begins to shake violently, announcing that a major earthquake is in progress. Instinctively we flee down the narrow, winding stairs from the loft, the distorted view provided by the reading glasses still perched on our noses impeding our progress almost as much as the madly rocking environment. We go right through the screen door onto the front porch without bothering to open it first, thus ourselves inflicting, as it turned out, the only damage our house was to suffer.

When the ground and trees had stopped shaking, I remembered Max and rushed back into the laundry room. The top of the washer was littered with containers of soap and bleach and cans of polish that had fallen from the open shelves above, but no Max was to be seen. I finally found him cowering in the deep laundry tub of the adjoining bath, unhurt but pitiful to behold. He was sharing the tub with bandaids, sunscreen, brushes, tweezers, a bottle of mercurochrome — objects from the medicine cabinet, some of which had no doubt fallen on him after his arrival at what he had probably viewed as a safe haven.

Poor Max. Not only was George gone, but his favorite room — the only room in the house in which he and George were allowed, the room in which they were fed, the room in which they curled up together in their basket — had somehow turned violently against him. This was even worse than the snowstorm. Everything was wrong with his world and he was glad to be rescued. I picked him up gently and soothed him with long, slow strokes while carrying him outside to what we all decided was a safer environment. It was a long time before he could be enticed back into the laundry room, even to eat.

Miraculously, our power, which is strung from tree to tree a long distance through the woods, never failed. As the aftershocks continued, we stood outside on the deck and craned our necks to peer through open doors at a built-in television set in the library. Thus we learned, after a short period of initial blackness, that the epicenter was nearby in the Santa Cruz mountains, where the damage, especially to older buildings, was horrendous. We no longer resented the hundred or so expensive pilings the county code had forced us to provide as a foundation, one that had seemed more suitable for a skyscraper than for our modest cottage.

Two days later George reappeared, to everyone's particular relief because this had been his longest absence yet. As usual he was neither tired, nor dirty, nor hungry, nor especially friendly — just home from wherever it is that he goes. And where had he been during the crucial seconds? Had he been high in a tree, hanging on desperately while it swayed back and forth? Had he been chasing some small rodent, missed his aim, and wondered how he could have miscalculated so badly?

I wondered if his travels had been prompted by a premonition of the 7.1 shaking we had just endured. I knew that animal shelters often reported excess wandering in advance of earthquakes, and many animals seem capable of detecting and being frightened by the minor tremors, magnetic upheavals, radon gas, static electricity, or extra low frequency magnetic fields (pick your favorite theory) that are often precursors. We had observed this ourselves some years ago when our finches, sleeping peacefully on their perch, awakened suddenly and started fluttering violently about in their cage, which was suspended by a thin wire from the ceiling. I'd had just enough time to say, "What on earth can be the matter with those finches?" when the answer became apparent in the form of a sharp jolt from a 5.5 earthquake. The next time George disappears, we'll probably take the most treasured items out of the china cabinet and store them more safely until he returns.

Deep in the heart of the stacks

The earthquake, as practically the whole world knew, had severed the San Francisco Bay Bridge, making access to my alma mater very difficult. I began to consider alternatives.

It wasn't loyalty to my alma mater, however, that had kept me in the past from patronizing the splendid libraries of its arch-rival; it was the price. The main reference library of the opposition was open to all, as long as hundreds of dollars were proffered. Otherwise one could enter only a few times a year, and no materials could be withdrawn. It was a gloomy prospect.

But the specialized libraries turned out to be much friendlier. I still couldn't take anything home, but I could come and go as I pleased. And I pleased a lot, especially on weekends when they were nearly deserted and parking was not a problem. Most journal articles were short, so I could afford the copying fees and could peruse them at my leisure. And there didn't seem to be any mad librarians. Things were looking up.

The open stacks provided access not only to the most recent journals in genetics and biology but also to ancient, dusty copies, with elegantly engraved illustrations, from around the turn of the century. These less current items, however, were bound together to form weighty tomes and hidden away in moveable stacks that were daunting to enter. Despite the electronic safeguards, I sometimes imagined being crushed by these devices as they closed ranks in their space-saving operations.

Since the earthquake, the movable stacks were even more frightening than usual. Whenever I entered, I visualized being buried by toppling books as the earth decided to readjust its plates yet once again. Even worse, misalignment from the recent shaking now caused the tops of the moving stacks to brush the fluorescent light fixtures dangling overhead. Sensing this obstacle, the stacks, with their heavy burden of books, would rock noisily back and forth, refusing to lock properly into position. I would dart in and out like a magpie, hastily grab a heavy volume, and then make a dash for safety and the copier.

My persistence was rewarded. I found that shortly after the rediscovery of Mendel, a series of papers had been published in which various attempts were made to explain the appearance of calico cats, especially males, in terms of Mendelian factors. By chasing the references at the end of each such paper that I located, and dashing quickly in and out of the quivering stacks, I soon had a fairly complete picture of the cross-talk that took place among these pioneers of cat genetics between 1904 and 1932.

And cross talk it was, too. They were vexed with intractable problems, with each other, and with the cats, whose behavior certainly left much to be desired. While famous figures in the field like Morgan and Sturtevant were happily describing the virtues of their wonderful flies, the little-known cat people were unhappily bemoaning the difficulties of dealing with their intransigent cats.

Doncaster leads the way

Darwin and Mivart were of the opinion that male calicos look quite different from their female counterparts: Darwin says they are "rusty-red," Mivart that they are "sandy-coloured." Both puzzled over why the male calico should have a color scheme all his own, but it's clear from their writings that neither of them had ever seen one — they were describing a cat of a different color.

Bateson's disciple Doncaster was also puzzled about why certain variations are confined to one sex alone. He had observed special coloration in the female moths with which he was working, and had also noted that human color blindness usually afflicts males only. He decided that maybe he could learn something about these general matters of sex-limited inheritance by trying to identify the specific Mendelian factors of the calico cat.

By the time he wrote his first serious cat paper in 1904, the calico situation had changed considerably from Darwin's day. Doncaster starts by reminding us that "it is commonly said that the corresponding colour in males is orange (otherwise described as red or yellow)," a belief he himself had held in earlier days. But he goes on to describe a mating, brought to his attention by a breeder, between a pair of male and female calicos that produced calico kittens as well as orange ones and black ones. He then asks why calicos are "almost exclusively females, the number of certainly known males of this colour being very small?"

Doncaster observed that when you cross orange cats with black ones, it matters what their sexes are. Orange mothers and black fathers give calico females and orange males (as shown ever so long ago in Figure 1), whereas black mothers and orange fathers give calico females and black males (as shown in Figure 2). Put generally for orange and black, the female kittens are all calico, and the male kittens all acquire the color of their mother.

To account for this curious state of affairs, Doncaster proposes what was to be the first of many theories about calico cats. He starts with the problem of how their color scheme arises and assumes that there is a pair of factors, orange and black, that compete at the same location. Then, expanding on Mendel's description of dominance, he invents something new: incomplete dominance — and sex-dependent at that. He suggests that in the male orange is completely dominant over black, while in the female the dominance is incomplete, and tortoiseshell results.

He's happy about his theory because it "accounts also for the fact that orange females are very rare, although males are common." (This was probably true in Doncaster's day, when breeders didn't realize that in order for a female to be orange, she must inherit an orange gene from both parents; otherwise she will be calico.) He's a little unhappy about his theory because no orange males — only black ones — are obtained from the mating of black mothers with orange fathers. His incomplete-dominance theory would predict their occurrence, but they just don't show up, and he has no explanation for their apparent absence. To deal with the problem of the occasional male calico, Doncaster theorizes that in this case, the usual male dominance of orange over black is out of order and calico results, as it would in a female.

And so things sat until 1912, when the American C.C. Little took a crack at the problem. His research interests were quite different: He was trying to understand the "sex-producing factor," about which the evidence was conflicting. In some species the females seemed to be XX and the males XY or X0 (the 0 representing an absence of one sex chromosome and consequently an uneven number of chromosomes in general); in others, such as birds and butterflies, it seemed that the females were XY and the males XX. He thought he could help sort out this muddle by studying the problem of the calico cats.

Little was particularly interested in the lack of orange males that Doncaster mentioned and wanted to see for himself what happened, so he crossed four black females with the same orange male. Like Doncaster's breeders, however, he failed to obtain any orange male kittens. His next step was to see what happened when he crossed a calico female with an orange male (this was the exercise left for the reader just after Figure 3). He reports (correctly) that he expected to obtain calico females, orange females, black males, and orange males. But having avoided breeder problems, he has caretaker problems instead, and more experimental data will be needed:

> One litter had been obtained from this cross; it contained one tortoise female, one black male and three yellows (dead), the sex of which was unfortunately undetermined before the caretaker discarded them.

Little believed the "black coat color in cats to be linked with the X element, and therefore to be sex limited," a situation already observed in *Drosophila*. As for the male calicos, he assumes that because they're so rare, they must be due to some "distinct mutation" — presumably a calico gene? — rather than to conflicts between black and orange (or yellow, as he and others insisted on calling it).

Doncaster responds in print immediately, saying that by now he has evidence "from a breeder who is thoroughly reliable" that sometimes black females occur where they're not expected — from crossing black females and orange males, and also from crossing calico females and orange males. Along the lines of his earlier theory of incomplete dominance, he suggests that perhaps sex-limitation is not absolute but partial. Differing with Little, he now thinks that maybe orange is sex-linked, but not black... .

By 1913 Doncaster writes another paper, presenting more evidence on sex-limited inheritance. He's found it not only in moths but also in chickens, canaries, and pigeons; for humans, he's looked into data on hemophilia, color blindness, night blindness, and nystagmus (the rapid and uncontrollable oscillation of the eyeballs). But he reports that these pedigrees are not reliable, especially inasmuch as the female carriers of those conditions are not visibly marked in any way. If they don't have sons who manifest the disease, there's no knowing what their genetic constitution might be. Calico cats, however, flaunt their genes and are thus much better subjects to work with. Perhaps they will provide helpful clues to the method of transmission of these human diseases and hence also to an understanding of sex determination.

And there are now more examples of male calicos. Doncaster has managed to acquire one for himself and is happily starting to plan his own breeding experiments, relieved at no longer having to rely on data from others. He decries the fact that the breeders who own these rare males mate them only with calico females in an effort to perpetuate the line (I know of a vet who is still trying to do this today). Instead, Doncaster wants to mate them with black females to test some of his linkage theories. He thinks now that black and orange may both be sex-linked, but not at the same location... .

The next year we get the bad news — his precious male calico is sterile. He "has mated, apparently successfully, with each of four females several times, but none of them have become pregnant." Doncaster reports on at least three other male calicos who are sterile and says that "there are hardly any records of offspring of tortoiseshell males, and the few that exist are perhaps not wholly above suspicion."

The focus of this paper is not on sex-limited inheritance but on the causes of sterility. Doncaster proposes that

> the rare tortoiseshell male is produced only by the abnormal transmission from the sire to a male child of a character which normally goes into female producing gametes ...

and then concludes that

> when, by failure of sex-limited transmission, an individual arises which receives from one parent a character which it normally receives only from the other, that individual tends to be sterile.

By 1915, Doncaster has removed one of the testicles from his disappointing tortie tom and reports that it looks normal but contains no seminal fluid — no trace of spermatogenesis. Yet his "sexual instincts were exceptionally strongly developed." The puzzle now is what causes his sterility. Comparisons are made with undescended testes in men and dogs, the question being whether the testes failed to descend because they were abnormal or were abnormal because they failed to descend. When the conclusion is reached that sterility is the result of retention within the abdomen, a different theory is needed for the tortie tom, whose testicles were fully descended. Can the problem be its female coloration, as he suggested earlier? Are all male calicos sterile?

Doncaster checks all the records of the Cat Fancy and fails to find a "single case in which a ... tricolour tom is recorded as a sire," although these cats are much prized by fanciers. But the Baronet Sir Claud Alexander (whose word is apparently wholly above suspicion) has had five male calicos, and the one named Samson was "undoubtedly fertile; he sired many kittens by tortoiseshell dams, but produced no tortoiseshell males." Doncaster now believes in the existence of at least one fertile male, but he wonders in general whether the color causes the sterility or the sterility causes the color. Despite the need to explain the few exceptions like Samson, he opts for the former, maintaining his theory of 1914 that possessing "characters proper to the female" is what's causing the problem.

Others get various words in edgewise, all using their own specialized notation. Most describe complicated interrelations of many genes, at many locations, and believe that there may be a calico gene at work along with two different types of black. But by 1919 Little is back in the fray again, trying to straighten things out. He enumerates the many problems with special clarity: nonreciprocal results on mating; unexpected results (such as black females); practically no males; the males usually being sterile; when not sterile, males breeding as orange. Then he points out that "investigators have usually tried to explain all of them by a single hypothesis." Because this hasn't been successful, he postulates "two genetically independent agents at work in the production of these aberrances" and goes on to lay out a very complicated scheme in which black and orange are independent; this is hard to follow but seems to take care of all the observed exceptions.

As far as the sterility is concerned, Little suggests (almost correctly) that it's caused by non-disjunction of the X-chromosomes, pointing out that this was observed in *Drosophila* in 1916. Little "places cats in the same category with *Drosophila*'" and suggests that "one cannot fail to be impressed by the similarity between the results of that process in *Drosophila*, and the observed experimental facts in cats."

He's right about the similarity, but only up to a point. In both cats and flies, the first step is indeed the reluctance of the two X-chromosomes of a female to part during meiosis; this results in one egg cell with both Xs still together and another egg cell with no Xs whatsoever. In flies, however, it's the sex-free egg that is fertilized by an X-bearing sperm to produce a sterile male of X0 constitution. In cats, it's the egg with the two Xs that is fertilized by a Y-bearing sperm to produce a sterile male of XXY constitution. Close, but no cigar.

Doncaster doesn't buy it anyway, pointing out that "the flies are almost always mosaics of sex-characters," whereas there is no evidence that this should be true of an X0 cat. In 1920 he eschews reluctance but proposes a truly new and different theory to explain sterility: What about freemartins! It's been known since at least 1681 (when the first reference to this term is cited in the *Oxford English Dictionary*) that if a cow bears two calves of different sexes, the female fetus is often "masculinized by the confluence of its vascular system with that of a neighboring male foetus." The result is called a freemartin (for unknown reasons, although *mart* is the Gaelic word for "heifer"). Maybe that's what's going on with the cats, and the male calico is really a female in disguise. Doncaster acknowledges that females of all color schemes would be similarly affected but points out that most would escape detection.

The advantage of his new theory is that it should be easy to test. Doncaster hastens to add, however, that he "can't undertake this considerable labour" but hopes that "some others may be able to obtain and examine the necessary material." Little throws cold water on this suggestion, saying that his own hypothesis is just as likely to be correct. But one of his students takes Doncaster up on it and tests 653 embryonic kittens from the uteri of 148 mother cats; she finds none of their vascular systems to be conjoined.

As Doncaster's colleague Mrs. Bisbee reports in 1922, Doncaster himself

> began an examination of all pregnant female cats available. He had examined fourteen when he died, and I have continued his observations up to the present time. Altogether 70 cats have been examined, giving a total of 253 kittens, and so far no case of confluence of blood vessels has been found. ... In one there was a slight attachment of the chorions of two adjacent embryoes, but unfortunately it was impossible to settle definitely by injection whether or not the blood vessels were confluent, for by an accident the kittens were moved in my absence and the two had separated.

(If she and Little hadn't been working on different continents, one would suspect the same caretaker, bustling about, keeping the lab nice and tidy.)

So the freemartin theory seems to have been laid to rest along with Doncaster, although in 1928 someone reports that "in opening up a cat last February ... what looked like complete fusion of placentae was seen" and suggests that Doncaster might have been on the right track after all.

Mrs. Bisbee carries on

But Mrs. Bisbee, back in 1922, has another idea. Following in her mentor's footsteps, she suggests that

> the male physiology may be favourable to the dominance of yellow over black, and the female physiology not so favourable. If the colour be a matter of sex physiology then by castrating a very young yellow male and grafting ovaries it might be possible to bring up the black to some extent later. Similarly by grafting a functional testis into a newly born tortoiseshell male it might be possible to inhibit the development of black in future coats. ... Administration of extracts of the endocrine glands and transfusion of blood might also give interesting results. I hope to attack the problem along these lines in the near future.

She adds that she's "had the opportunity of dissecting Professor Doncaster's tortoiseshell tom cat" and found his second testis nonfunctional, just like his first. She says that she still hopes to try the matings with black females that Doncaster looked forward to "if ever I am fortunate enough to find a fertile tortoiseshell male."

In 1927 Mrs. Bisbee reports that she did indeed castrate three newly born yellow male kittens and proceeded to feed two of them with ovarian extract for six months, but "the results were entirely negative" (I'm sure the kittens would concur). She had intended to do the same thing to some black male kittens, but "our genetical work practically proved that there is no difference in dominance in the two sexes, and our physiological work was therefore discontinued."

This kitten-saving proof was acquired because of fleas. In discussing the difficulties of even determining the colors of the cats correctly, Mrs. Bisbee explains that

> The yellow female was at first thought to be an ordinary yellow kitten, but when she was about four months old three minute black spots were discovered on the back of her right hind foot. She was then, of course, recorded as a tortoiseshell.
>
> ... After the discovery of the black spots ... we naturally examined all our yellow cats very carefully, but none had the least suggestion of a black spot anywhere. Later one of our yellow kittens became infested with fleas and when going over it very carefully with a small tooth comb, we found three or four black hairs. No ordinary examination would have revealed this small amount of black and consequently we began to examine every available yellow cat with especial care. We have examined cats from our own stock, from the Cats' Home here in Liverpool, from different parts of England and from the Isle of Man and the interesting fact has come to light that apparently *all* yellow cats have a few black hairs ... .

Mrs. Bisbee uses the fact that black hairs have been found in all yellow cats of either sex to "practically disprove any sex-difference in the dominance of black and yellow." After reviewing, yet again, all the current complicated theories, Mrs. Bisbee proposes a new one: "there has been fractionation of a factor." She suggests that the only difficulty in the way of acceptance of this theory is "the deeply rooted but purely hypothetical conception of an indivisible gene."

In 1927 Mrs. Bisbee publishes the exciting news that her wishes have been granted: She has finally acquired, from a Mrs. Langdale of the Cat Fancy, a fertile male tortoiseshell named Lucifer — and he's not just any old tortie tom but a tabby tortie tom (actually a torbie-and-white, like George). These torbies, she says, are thought to be extremely rare, and only one other has been definitely recorded. (She's probably referring to the one featured at the Crystal Palace Cat Show of 1912.) It's too early to report on his progeny as yet, but she embeds this announcement in a short note whose main purpose is to discredit the opposition. They've suggested a special gene — dominant black — as the key to the calico puzzle and have announced that they're in the possession of *two* male calicos that were born in the same litter, the result of mating a yellow male with a Siamese female. She writes,

> In view of the extreme difficulty of sexing some kittens at birth it will not be considered unduly critical if we express a hope that these recorded tortoiseshell males will either be allowed to grow up, or be dissected. ... they *may* have been sexed incorrectly. We dare to suggest this only because the occurrence of *two* males in one litter both inheriting yellow from their father is such a very extraordinary result.

The scientists under attack, however, have their data under control and are able to reply as follows:

> The criticism by Mrs. Bisbee ... was rather welcome to us, since, in our reply to it, it gives us an opportunity of publishing some more details on our tortoiseshell males. ... We regret to state that one of them died when only two days old. He was however dissected and proved to be an undubitable male. The other one is still alive. He is now 17 months old but so far has produced no offspring, although several queens were offered to him. Most prob-

ably he has never copulated, and we are forced to suppose that he is abnormal, or at any rate infertile.

In 1931, still following in Doncaster's footsteps, Mrs. Bisbee tries to ascertain just how sterile male calicos really are. Looking into the scanty available statistics, she reports that up to 1915, only seven had been recorded, whereas in the intervening sixteen years, another seven have been found. Of these fourteen specimens, there is what she considers firm evidence on only eight: three of these were fertile, four were sterile, and the last one was probably sterile as well. (The fertile cats have names like Samson, Lucifer, and King Saul; the infertile ones have names like Bachelor and Benedict.) She writes that "it does appear ... that the abnormal association of black and yellow in the male cat is correlated with a tendency towards sterility," explaining that she bases this conclusion on the fact that

> there are no published records of the incidence of sterility among ordinary male cats, but in the course of our own breeding experiments we have used fourteen such males, taken at random from the general population, and all have been fertile.

In 1932, Mrs. Bisbee writes a sad last paper on the topic. It is in memory of Lucifer,

> the only male of this type which has hitherto been available for purely scientific breeding experiments. Unfortunately he has recently died; therefore, although our experiments are not completed, there is no reason for further delaying the publication of our results.

As Doncaster wished, she has mated Lucifer to black females (and also to orange and calico ones). She carefully records the colors and sexes of the 56 kittens he thus fathered (perhaps she wore him out?) and concludes that indeed he breeds just as if he were orange. She reports that his daughters gave normal results when mated to unrelated males, but that unfortunately we are not able to give any data in regard to the offspring of his sons. A few were chloroformed when newly born and some died before reaching maturity.

She reviews yet once more the many theories that still abound and this time opts for a new one, again involving fragmentation: She calls it partial non-disjunction. Lucifer's mother was known to be a tortoiseshell, and she suggests that

> if, in the formation of the gametes of this tortoiseshell female, *part* of an X-chromosome, carrying black, failed to separate from the yellow-carrying X-chromosome, she might well produce a tortoiseshell son.

And that's the last of Mrs. Bisbee's theories about the male calicos. In an enormous review which she wrote for *Bibliographica Genetica* in 1927, covering all the calico cat work up to 1924, Mrs. Bisbee asserts that

> some of the most interesting problems in Genetics are connected with the inheritance of the black, yellow and tortoiseshell coat-colour amongst cats.

(She's highly prejudiced, of course.) She explains that when Doncaster died he left her all his notes, letters, and records about the cats, but still the problem of the male calicos and the unexpected black females is by no means settled; even the normal mode of inheritance of black, yellow, and tortoiseshell is not well understood.

The library can wait

My uncle has Alzheimer's disease. He's been visibly afflicted for five or six years now, but who knows how long he suffered before that, living alone as a perennial bachelor, hiding from himself and others his diminishing ability to remember recent events or to find his way around in the city he's inhabited for the past half-century. His daytime nurse just called to say that she can hardly get him in and out of a taxi anymore and that a van with a wheelchair lift will be necessary if he's to escape the confines of his small apartment. Slowly, but very surely, his brain is filling with the plaques and tangles first described by Dr. Alzheimer in 1901. In this manner, my uncle is inexorably approaching the fate of his father, who was similarly afflicted and spent the last seven years of his life lying in a hospital bed, unable to recognize anyone.

My mother too is starting to struggle with her short-term memory, a not uncommon problem for people in their mid-eighties. This doesn't mean, of course, that she has Alzheimer's, but it makes me wonder a little uneasily whether we're watching genetics in action. Rather than hiding her difficulties, however, my mother remarks upon them and recently sent me a long clipping from *The New York Times* (February 6, 1991) describing various theories about possible genetic and environmental causes of Alzheimer's.

The story was prompted by an article in *Nature* reporting a suspect mutation on chromosome 21. There have been earlier theories about troubles on chromosome 19. But there are also large Alzheimer's families whose afflicted members don't seem to have unusual genes at either of these locations. Is there a third bad gene yet to be discovered? Do these genes work in combination with one another? Or are the causes instead environmental, either wholly or in part? Is Alzheimer's, which is diagnosed only by the appearance of plaques and tangles found in the brain at autopsy, actually a catch-all term describing a set of different diseases with different causes?

I think I should look up the *Nature* article, should keep informed of the latest developments in an area where I have been (and may again be) so personally affected. But then I recognize the surprising similarities between this enterprise and the one I've been engaged in all week: reviewing the early papers about the genetics of calico cats. I've found these papers fascinating, pitiful, hilarious, frustrating, and inconclusive. Will the offerings in the Alzheimer arena, reviewed by some counterpart fifty years hence, evoke different responses?

Probably not. Although modern geneticists have access to cytological tools and information that the early cat people couldn't have imagined, the situation of the Alzheimer's researchers is not so very different from theirs: The muddle is large, the data are squishy, and the conjectures are all over the map. Sorting out Alzheimer's may take a very long time. Perhaps not as long as the sixty odd years it took to determine the genetics of the male calico, given that the motivations and rewards, and the foundation of knowledge, are all much greater. Still I don't think I'll rush to the library. As with the cats, there will be many more rival theories, hotly contested in the appropriate journals, and much theorizing, attacking, and retracting before this problem also is considered solved and laid to rest. Viewed in retrospect, some of these current Alzheimer's articles may seem almost as quaint as those the cat people wrote during the early decades of this century.

Isaac Newton would be amazed

More than a year has passed since we trapped Oscar, and the nights have been pretty peaceful around here — until a few weeks ago, that is, when we acquired a new case of the 4 A.M. blues. The cause was Oscar's successor, yet another fierce fighter of a cat who also knew a good thing when he saw it and decided to lay claim to our territory. We might have named him The Phantom (he appears charcoal black from a distance but leaves long silvery clumps of hair to mark the scenes of various battlegrounds), but instead he's been dubbed Houdini in recognition of his amazing ability to elude our determined attempts at entrapment.

We've had three opportunities to view this interloper at close range: once when my husband tried to put him into a box and Houdini leapt out of his gloved hands; once when I tried to put him into a box and he made a deep gash in my ungloved little finger; and this morning when we were at last triumphant and managed to catch him in the laundry room.

Close up, one can see that Houdini's fur is not black but silver (or chinchilla, as the cat fanciers like to say). Only the very tips of the hairs are darkly pigmented, but this is enough to make him seem dark all over — until he bends or scratches and exposes all that silver lining. This wonderful effect is caused by the dominant mutant gene I, which inhibits the development of pigment and produces pale hairs that are "tipped" with varying amounts of color, depending on how many inhibitor genes are present. Now that he's in our clutches at last, we can see that Houdini also has dark brown tabby stripes that make pleasing but barely discernible patterns across his face and body. Our dark phantom is actually a silver tabby, an elegant creature who has somehow fallen on hard times.

Like the Oscar of the past, Houdini is hungry, lonely, and obviously accustomed to the better things of life. We're trying hard to provide them for him and have found a nearby artist colony that's eager for his company and services. But Houdini is also tough, canny, mistrustful, and self-sufficient — as with Oscar, we can't make him understand how wonderful his life is going to be just as soon as he gets into that box. To complete the similarities, Houdini is also partial to 4 A.M. as the best time to engage in a noisy fight with George.

After a week of sleeplessness, we rented what looked like a safe, clever, and splendid trap. Houdini agreed. He happily came and went, taking whatever offerings were provided, including a chicken bone wired tightly to the platform he was supposed to step on to cause incarceration. We have no idea how he goes about it; sometimes the trap is sprung, sometimes not, but he always manages to make off with the booty.

After a week of paying for the trap, we admitted defeat and returned it, weary from lack of sleep. George was showing wear and tear as well. First it was a scratched face, then a deeply gouged front paw (like my little finger), then a bite out of the back of his hind leg, and finally some desecration of his shrunken scrotum. Max, as usual, remained unscathed. (He supervises like a referee from some safe, high vantage point and then solicitously licks George's wounds and prepares him for the next encounter.) The raids continued with no end in sight. What were we to do?

Puttering in the kitchen one evening just at dusk, I heard George or Max being unusually noisy with their food dish in the laundry room. I went to see which one was exhibiting such bad manners and arrived just in time to catch a glimpse of Houdini's hind legs disappearing out our fancy new high-tech cat door.

Isaac Newton is said to have invented the cat door, but even he would be amazed at how his design has been modified. This modern version can flap in either direction (like Newton's) but is made of heavy plastic and latches firmly into position when closed. It's also airtight, watertight, and raccoon-tight — in fact, it allows only those sporting an appropriate electronic widget to enter. Half a mile away, Oscar has a widget and a high-tech cat door of his own. But the widgets come in various versions, and we've been careful to choose one that's different from his. So although Oscar would presumably recognize our cat door, our cat door wouldn't recognize Oscar's widget.

So how did Houdini manage to enter? The answer is quite simple — the door wasn't plugged in. This omission was not an accident but a deliberate concession to Max's fear and loathing. Although George learned quickly how to use this magic door and clearly appreciated the freedom of access it provided, Max just as clearly preferred the old method of interaction in which we leapt up to open the back door for him whenever he wanted to come or go. After weeks of trying to teach him just to push the clear plastic flap open with his nose (by kneeling on the other side of it with our rumps in the air, waving goodies in his direction), we finally decided it was time to activate the door by plugging it in. But when he heard the slight buzz the door made as it recognized the widget on his collar, Max leapt back in horror. All our weeks of progress were lost as he made clear to us that nothing would persuade him to go near this infernal contraption ever again. We finally gave up, pulled the plug, and resumed our previous lessons, hoping that the inactivated but still sturdy cat flap would provide sufficient protection.

(Max has had similar reactions to the long hose of the central vacuum cleaner. Once I forgot to check their basket before cleaning the laundry room and found Max arched in fear as I approached. He seemed to be thinking, "Boa constrictor! Anaconda! Burmese python!" in rapid succession but didn't know what to do about any of them. Remembering a newspaper story about a man who had ended up in the hospital with hundreds of stitches because he had started up a powerful new vacuum cleaner in the presence of his terrified cat, I unplugged the hose and put Max gently outside before proceeding.)

Houdini's invasion made us realize that the laundry room itself was the trap we needed. Even without electrification, our fancy cat door can be set to open in neither direction, in either direction, or in both directions via a sturdy rotary switch. We arranged a beautiful dinner for Houdini and set the switch carefully to allow ingress but not egress. Then, not wanting George or Max to be trapped with Houdini, and not wanting to lock them in the cold garage either, we took them with us into the forbidden territory of the house.

Max was ecstatic and made straight for the window seat in the library. (He remembered this favorite spot from a week several years ago when he'd been under treatment for a severely scratched eye.) George was inquisitive and investigated the new terrain very carefully, both upstairs and down, before settling happily into the box of computer paper under the desk in the loft. We also went to bed very happily, expecting to sleep through the night for the first time in weeks and to find Houdini waiting for us in the laundry room in the morning.

But our expectations were not to be fulfilled. After polishing off his dinner, Houdini had lived up to his name and escaped, though not without considerable difficulty. By persistent clawing at the switch, he had finally managed to twist it into the position needed for his release. We couldn't believe it. His I.Q., we decided, must be at least 150.

He was certainly outsmarting us, but we felt sure he'd return, and this time we'd be ready. My husband carved a small piece of wood that wedged the switch tightly into place, and we removed absolutely everything from the laundry room, imagining the havoc Houdini might wreak with curtains, cleanser, soap, and bleach when he realized that this time he was trapped for good. Another tempting feast was arranged, and George and Max were led back into the house for what we hoped would be the last time.

This morning a much-chastened Houdini was to be seen sitting on the window sill mewing pitifully. The cat door and its immobilized switch had been violently clawed, but to no avail. The curtain rod was hanging askew, the screen we'd forgotten to remove was on the floor, the window had been thoroughly sprayed, and the place had acquired a Houdini-smell that may be with us forever. We called a writer from the artist colony who very skillfully, by dint of much soothing conversation in cat language, followed by a tenacious half-Nelson, persuaded Houdini to get into his car for a trip to the vet where tests, shots, and castration awaited.

The Houdini saga is over and we can sleep again, at least until the next abandoned cat decides to fight it out with George for control of the territory. This time, however, the residue is more lasting than it was with Oscar. Even after a vigorous scrubbing the smell remains intense, and we've filled the laundry room with hyacinths in retaliation. George and Max aren't fooled, however; they feel sure Houdini still lurks within. They won't use the cat door and will enter their former domain only when we open the back door for them — and then only with great wariness.

George's face is starting to swell where Houdini scratched and bit him, and tomorrow's agenda will no doubt include the lancing and draining of yet another abscess. But Max is the one who's sulky and depressed. He parades around the decks, pressing his nose against every available piece of glass. He peers into all the rooms to see what we're up to, to make us feel guilty, and to let us know that he doesn't understand his fall from grace. What has he done wrong? Why is he no longer allowed access to the Promised Land?

8 *Twentieth-Century Theories of Sex*

Mary Lyon's mottled mice

AFTER LUCIFER'S DEATH IN 1932, THE PROBLEM OF THE MALE CALICO seems to have been temporarily abandoned: I find only a single paper, from 1941, which makes any attempt to deal with it during the next twenty years. Here the research focuses on the study of meiosis taking place in the testes of tabby, black, and yellow cats in the hope of shedding "some light upon the highly complex genetical behaviour of tortoise-shell cats." The Scottish researcher reports that the "X and Y sex chromosomes are very similar in size" (the X-chromosome of the cat wasn't correctly identified until 1965) and behave strangely in a number of different ways; he rather lamely concludes that the tortoiseshell males can be accounted for through "structural peculiarities of the sex chromosomes, particularly those of the Y."

In 1949, however, an important genetic breakthrough occurs in which cats are involved (although not calico ones as far as I know). Two Canadians named Barr and Bertram publish a one-page paper in *Nature* announcing that female cats can be reliably distinguished from male cats by the presence of a dark blob, easily seen within the nucleus of their cells. Similar dark blobs — or Barr bodies, as they came to be called — are next found in the cells of female rats, mice, Chinese hamsters, and humans. In fact, they are found in all normal female mammals tested, but they're never found in normal males. No one knows then what these Barr bodies are about, but they're clearly connected with sex in some manner.

More detailed investigations show that the two Xs of normal females never behave alike: One replicates itself just like all the other chromosomes; the other is late in replicating and has a Barr body. But like most interesting results, this one leads to yet more questions: How is it determined which of the two female Xs has the Barr body? What is its function? At what point in development does it appear?

Answers to these questions were to crack the calico cat problem wide open (and a lot of other important problems as well). They were provided in another one-pager entitled simply "Gene Action in the X-Chromosome of the Mouse (*Mus musculus* L.)," which was written by Mary Lyon and published in *Nature* in 1961. You might think that finding such a short paper with such a seemingly irrelevant title would be a difficult problem. Not at all. This is one of the seminal papers in genetics, and references to it are abundant.

In essence, Lyon proposes the following simple but remarkable hypothesis:

- Barr bodies indicate the inactivation of the X in which they are seen.

She fleshes this out by observing that

- In female mammals, it's normal for one of the two Xs to be inactive.

- Random inactivation occurs very early in embryonic development.

- In some cells the X inherited from the mother is inactivated, in others the X inherited from the father.

- Such inactivation is irreversible.

- It is copied in all cells descended from the original few.

- Thus all female mammals have two separate cell lines; they are all genetic mosaics where their X-chromosomes are concerned.

(It's important to note that the egg-producing germ cells are exempt from this process. They contain two active Xs, either one of which may find its way into an egg in good working order. It's also important to note that Barr bodies are not really separable bodies at all but merely the X-chromosome itself in a highly condensed, and hence inactivated, state.)

As support for what came to be known as the Lyon Hypothesis, she cites her mottled female mice. Their variant coat colors were known to be caused by genes on their X-chromosomes, and she suggests that their mosaic appearance is the result of the random inactivation of different Xs in different cells. Thus patches of color A may be caused by cells descended from those in which the maternal X is in charge; patches of color B may be caused by descendants of those in which the paternal X is calling the shots instead. (The male mice, having only one X to work with, are not mottled.) Although her evidence is based on mice, she concludes with a note about cats, reminding us that "the coat of the tortoiseshell cat, being a mosaic of the black and yellow colours ... fulfils this expectation."

Her hypothesis was rejected in 1967 by a Professor Grüneberg after "a searching critical evaluation." But many new facts, from many different sources, continued to emerge in support of her hypothesis, and eventually Mary Lyon triumphed. In 1974 she dedicated her long review lecture on this topic to Dr. Grüneberg, on the occasion of his retirement as Chair of Animal Genetics, University College London. The Lyon Hypothesis was no longer merely an hypothesis; it has become an accepted fact of life with many interesting ramifications.

A trivial but amusing one, pointed out by Irene Elia in her book *The Female Animal*, is that girl identical twins are almost always less identical than boy identical twins. Although in both cases their genetic material originates from a single fertilized egg that cleaves in two, the pair of female embryos will undergo random inactivation of their Xs about 10 to 12 days into their development. Because several thousand cells are involved, this random inactivation is quite unlikely to produce identical results.

A more important ramification is that we must modify once again our basic description of conflict resolution in which the pair of genes at a particular location is depicted as duking it out, the dominant one always winning. As we saw earlier in the section on intertwingling, this simplistic picture doesn't always apply: there's masking, in which a gene (such as *W* for overall whiteness) is so powerful that it obliterates the effects of genes at other locations as well; there's variable expression, in which a gene (such as *S* for white spotting) produces more or less of its effect depending on how many of them are present; and there are other mechanisms in which polygenes — many sets of genes at different locations — all interact together in small but complicated ways to produce a single result.

But where genes on the X-chromosome are concerned, there's often no conflict to be resolved. Male mammals have only one X, and its genes always prevail over those at equivalent locations because there aren't any. If a male cat has the orange gene *O*, he will be orange — no question, no conflict. If a female cat has both an orange gene and a non-orange one (is *Oo*), there's no conflict either. Cells with an active X containing the gene *O* specify orange hair; cells with an active X containing the gene *o* specify not orange hair but black, just as Mary Lyon surmised.

(The size and placement of the orange and black patches depend not only on the distribution of the orange gene *O* but also on the prevalence of the white spotting gene *S*; environmental conditions within the womb are also a contributing factor. The skin shows these orange and black patterns too, as I learned to my surprise the last time the vet shaved George to treat one his many fight-inflicted abscesses.)

For other kinds of genes on the X, whose job is to make various enzymes that control bodily functions, conflict will result in variable expression: The amount of enzyme produced will depend on the percentage of each type of gene remaining after the purge.

After all that effort by the cat people, it would have been nice if they, instead of a mouse person, had been responsible for this particular piece of enlightenment. But they had another chance coming. Although Mary Lyon had provided a persuasive description of why female mammals, but not male ones, might have calico-style coats, she didn't have anything to say about the rare exceptions. There were still a few Georges out there to be accounted for, and with the Lyon Hypothesis as a springboard, the cat people were soon back in action again.

Klinefelters all

But there were a few questions about exceptional people that had to be dealt with first. After all, why should cats be the only mammals whose sex chromosomes are reluctant to part? Not only are there XXY people, who are designated as having Klinefelter syndrome, but there are also XXY dogs, pigs, sheep, horses, goats, mice, and Chinese hamsters — Klinefelters all. And the number of human Klinefelters, based on the screening of newborn male infants, really surprised me: roughly one in every 500 males comes endowed with an extra X — or two or three or more!

The statistics for cats and other animals are harder to obtain, but they're certainly different. Whereas human Klinefelters are relatively common, feline Klinefelters are thought to be quite rare. A rough estimate for calicos, based on a sample of 17,000 such cats, predicts that only 1 in 3000 will be male, and the difficulty that's been experienced during the last century in locating such easily identifiable objects lends support to such a tiny number. XXY cats, of course, may be of any color, and the ones that aren't calico will probably go unnoticed. Still it seems clear that, for some unknown reason, cats are more clever than people at sorting out their sex chromosomes.

So what is Klinefelter syndrome and what do afflicted humans look like? The symptoms were first noticed (by a Dr. Klinefelter, of course) in 1941. The very first such patient that he saw was an eighteen-year-old black youth complaining that he had small womanly breasts. He also had a deep voice and a large penis but had no beard and very small testicles. He was treated with heavy doses of hormones, both to reduce the size of his breasts and to increase the size of his testicles, but there was no appreciable result in either direction. During the next six months, Klinefelter encountered eight similar patients with a variety of complaints. A seventeen-year-old white schoolboy said he'd been rejected by the Navy because his testicles were too small (it wasn't clear for what tasks the Navy thus found him unsuitable). Several were of low intelligence, some had very high-pitched voices, many didn't like their female-style breasts and wanted to get rid of them, and three married men in their early thirties came in complaining of sterility.

These patients, and the studies that Klinefelter and his boss Dr. Albright made of them, were first described in a paper published in the *Journal of Clinical Endocrinology* in 1942. As Klinefelter says retrospectively,

> This is actually another of Dr. Albright's diseases. He unselfishly allowed my name to come first in the list of authors; because of the length of the title ["Syndrome characterized by gynecomastia, aspermatogenesis without a-leydigism and increased excretion of follicle-stimulating hormone"] and the convenience of the eponym, it became known as Klinefelter's syndrome.

Somehow it sounds appropriate — perhaps because the *Kline*, despite the spelling, inadvertently underlines the small larynx, small breasts, and small testicles.

By 1945 others reported more patients, some of whom didn't show breast development but shared the small testes and sterility. A few were mentally retarded; some had no beards or generally scanty body hair; some had small larynxes and high voices; many were extraordinarily tall (like George). They were termed Klinefelters as well, although at that time it wasn't known what caused the syndrome or how it should be characterized.

In the 1940s and early 1950s, Klinefelters weren't thought of as

XXYs at all; no one knew what their genetic constitution might be. This isn't surprising, because the human chromosome count was then thought to be 48 (24 pairs), a belief that had been held erroneously since around 1920. And it wasn't even known whether all cells contained the same number of chromosomes or whether it was the presence or absence of an X or a Y that was sex-determining.

Not until 1956 was the correct human chromosome count of 46 (23 pairs) determined, thanks to new cytological techniques devised by an Indonesian and a Swede. The belief that humans had 48 chromosomes was so strong and had been held so long that Tjio and Levan declare themselves surprised by their "very unexpected finding." They also mention in passing that a study undertaken by others in the previous year "was temporarily discontinued because the workers were unable to find all the 48 human chromosomes in their material; as a matter of fact, the number 46 was repeatedly counted in their slides."

These new techniques made it possible, at last, for accurate chromosome maps to be made. Chromosome pairs could now be distinguished from one another reliably, and by 1959 it became possible to see from their karyotypes that 80% of those termed Klinefelters had three sex chromosomes instead of two: They were truly XXYs and had a chromosome count of 47 (23 pairs plus one extra X) instead of 46. The remaining 20% had more extra Xs in various forms and combinations — and even some extra Ys as well. Figure 8 shows some of the abnormal combinations of sex chromosomes that have been found in humans and animals.

XXY XXXY XXXXY XXYY XX/XXY XY/XXY XY/XXXY

XXXY/XXXXY XXXYY XXY/XX XXY/XY XXY/XYY

XXY/XXXY XXX/XXXY XXXXY/XXXXXY/XXX XX(Y)

X0/XY/XXY XXXXYY/XXXXY/XXX Y0/XY/XXY XYY

X0/XX/XY/XXY/XXXY XY/XXY/XXXY XX/XY/XXY

Figure 8. Some abnormal sex chromosome combinations.

Mosaics and chimeras

A lot of the combinations in Figure 8 are tied together with slashes—these represent sexual mosaics. The word *mosaic* comes from the Greek *mouseios*, which means "belonging to the Muses, artistic." In biology, being a mosaic means having a mixture of different cell types within a single organism. There are, of course, mosaics in which other chromosomes are involved, but we'll just deal with sex chromosome mosaics here.

The first Klinefelter mosaic shown is XX/XXY, which happened to be the genetic structure of the first person with Klinefelter syndrome whose chromosomes were studied. This terminology indicates that some of his cells were standard female ones with two Xs and some were XXY cells having 47 chromosomes instead of 46 because of the extra X; it specifies nothing about the relative proportions of these two cell populations.

Klinefelter syndrome has a variety of causes, non-disjunction and the breakage and subsequent loss of chromosomes being most common. (As with Down syndrome, the probability of this occurring increases with the age of the mother.) In standard XXYs, the non-disjunction can occur at either the first or the second division of meiosis, when egg or sperm cells are being made by a parent of what will become a Klinefelter child (see Figure 5 to review the usual meiotic process, Figure 7 for the unusual Klinefelter result). In mosaics, the non-disjunction usually occurs during mitosis in the afflicted individual, often when the fertilized egg first starts practicing cell division and gets it a little bit wrong. Thereafter its division may improve, and it may proceed to copy both the "wrong" cells and the "right" cells faithfully, many millions of times. In this way, two or more different types of cells may exist in separate cell lines throughout the body.

Another way in which multiple cell lines can occur is through chimerism. In Greek mythology, the chimaira was a fire-breathing monster represented as having a lion's head, a goat's body, and a serpent's tail. In biology, a chimera is a special kind of mosaic that arises from the fusion of two embryos very early in their development. Famous examples of chimeras include moths that have the light-colored wings and thin antennas of the female on one side of their usually symmetrical bodies, and the brown wings and feathery antennas of the male on the other. A true hermaphrodite of this type — a fifty-fifty XX/XY mosaic — was first described with wonder in 1761.

Mistakes, of course, can be compounded, and it's not surprising that an organism with a tendency toward error will screw up more than once, perhaps in different ways. This accounts for very complicated mosaics, such as the one symbolized at the beginning of the bottom row of Figure 8, which has five different cell lines all jumbled together in the same individual.

Men with an extra Y rather than an extra X — XYYs — are not usually classified as Klinefelters. XYYs are thought to occur only half as often, at the rate of about 1 in every 1000 male births; they're usually even taller and larger than XXYs, may have severe acne, and are sometimes described as aggressive and violent. But this perception may be caused by their exceptional, seemingly threatening, size; or their behavior may be the result of mental retardation. Despite all the adverse publicity that has been circulated about them, only about 4% of XYYs ever find themselves in jail.

The most severely retarded and antisocial of those with various combinations of extra Xs and Ys are, however, locked up, away from public view. So in contrast to George, who advertises his genetic make-up by displaying black and orange blotches, most Klinefelters are hard to identify on casual observation. You'd never know it if you saw one on the street, although the statistics assure us that there are lots of Klinefelters on parade. Their most striking feature may be their height, but of course not all men with extraordinarily long legs are XXY; many of those who suffered from breast development have had their breasts removed for prophylactic as well as cosmetic reasons (their incidence of breast cancer is twenty times higher than that of normal males); many typical Klinefelters have normal intelligence (the degree of retardation seems to be related to the number of extra Xs, but only 20% of Klinefelters are more than XXY). And there are probably many who, like George, have no idea themselves that their sex chromosomes are so abundant.

How to make a male

The problem of sex determination has been of interest for a very long time. Not only do people have a natural curiosity about how it all works, but they've sometimes had a desire to influence the outcome. Hence many of the descriptions are given, even today, in terms of how to make a male.

The ancient Greeks put forth a multitude of theories (many involving heat), some of which persisted into the Middle Ages and beyond. Pick your favorites from the following list and try to imagine what it must have been like to put them into practice.

> The level of male excitement during intercourse determines whether the child will be male or female. The more excited the man, the greater the likelihood of siring a boy.

> The heat of the womb is decisive for sex determination: In a warm uterus, semen produces males; in a cold uterus, females.

> If the sperm is warm and present in large amounts, a boy will be born.

> An equal degree of heat in the semen of the parents gives rise to a boy who resembles his father. (The female egg was unknown and female semen was postulated.)

> The preponderance of male or female semen determines the sex of the child; equal amounts produce hermaphrodites.

> If the semen is strong and viscid, rather than weak and watery, the child is more likely to be male. (Hence it was thought that very young and very old men were more likely to produce females.)

> Copulation engaged in when northerly breezes are blowing produces males, southerly breezes females. (This is because the

north wind invigorates and thickens the semen, whereas the south wind enfeebles it and makes it watery.)

Ingesting hard, cold water causes infertility as well as the birth of females.

Right and left were associated with good and bad, male and female. The right side of both the testes and the uterus (which was thought to have two compartments, like the uteri of cats and cattle), was perfect, strong, and warm; the left imperfect, weak, and cold. Thus sperm from the right testicle, entering the right side of the uterus, produces males; left, left produces females; crosses produce hermaphrodites. This belief was still common in 1350, and women were counseled to lie on their right side after intercourse.

As late as the 1890s, many still thought that the penetration of more than one sperm was needed for conception and that the more sperm, the greater the likelihood of a male. Other theories abounded, some based on statistical studies, which declared that the sex of the offspring would be the same as that of the older or more vigorous parent, that underfed embryos turned out to be males, well-fed ones females. The list could go on and on.

Life in the XY Corral

There is undoubtedly current folklore just as bizarre as that of the Greeks and the Victorians, but at least the scientific community now has a fairly firm grip on what determines whether an embryo will be the kind that tries to make eggs or the kind that tries to make sperm. For the first six or seven weeks of its existence, the human embryo is "sex-indifferent" — males look just like females. Then suddenly a switch is flipped and the embryo decides to grow either ovaries or testes. Once that happens, all else follows: The now differentiated gonad makes either male or female hormones, and they take charge of the further stages of sexual development — a splendid example of pulling oneself up by one's own gonads.

Where mammals are concerned, it's been known since 1959 that the setting of the sex switch depends on the presence or absence of the Y-chromosome in the egg-penetrating sperm: If the Y is present, the resulting embryo will be male; If the Y is absent, the resulting embryo will be female. Studies of Klinefelters, both mice and men, were crucial to this new understanding.

A geneticist named David Page (who calls his lab "the XY Corral") claimed in 1987 to have pinpointed the exact tiny piece of the Y that was responsible for maleness by studying a few humans whose sex chromosomes didn't match their sexual machinery. Most important were a few "sex-reversed" males who showed up in doctors' offices complaining of sterility. They weren't XXY, as you might suspect from reading about the Klinefelters, but XX, like females — yet they appeared to be normal males in all respects except fertility. No one knew how they had managed to grow testes without any Y component in their genetic make-up.

Page theorized that something had gone wrong with their fathers' sperm-making machinery: A segment of DNA containing the male-determining gene must have broken off of a Y and attached itself to an X — "translocated" itself — probably in a rare case of the sex chromosomes crossing over during meiosis. Further, this X with the piece of Y embedded in it must have been on board the sperm that won the egg-fertilizing contest. (These sex-reversed males are represented as XX(Y) — as seen at the end of the third row of Figure 8 — the parenthetical Y being provided to indicate that there has to be some Y in there somewhere.) Only 1 in about 20,000 human males has no Y-chromosome, and rare accidents like these — and like George — help illuminate how things are supposed to work by providing examples of what happens when they don't.

There were also a few sex-reversed XY females around, and the commensurate theory was offered for them: Although they undoubtedly had a Y-chromosome, the piece containing the crucial male-determining gene must have broken off somehow and gotten lost. A few were found to have almost complete Y-chromosomes, and yet all were indisputably female. They weren't, however, as seemingly normal as their counterparts, the XX males: Some didn't exhibit female breast development; some failed to menstruate. So the theory that the mammalian embryo is basically female, being triggered into maleness only by the male-determining gene on the Y, needs some modification to account for these slightly deficient females.

To find the exact location of this gene, the trick was to find the smallest piece of DNA that the XX men held in common and to see if it matched the piece that all the XY females lacked on their Y-chromosome. Page studied about ninety sex-reversed individuals and had two pieces of very good luck: He found an XX male who had only 0.5% of a Y translocated onto an X, and he found an XY female who had 99.8% of a Y. This narrowed the search down to the tiny region held by the male but missing in the female. In all the other subjects, this region proved to be significant as well.

Having found what they thought was the gene, Page and his colleagues started searching for that exact same sequence of DNA in other mammals. And they found it everywhere they looked — in gorillas, monkeys, dogs, cattle, rabbits, horses, and goats — using what Page calls "Noah's Ark blots," paired male and female samples from every species. (Presumably, they would have found it in cats as well if they'd bothered to look, because there's strong evidence that mammalian sex chromosomes have diverged very little from one another over all these millenia of evolutionary change.) Finding this tiny sequence in all his samples confirmed to Page that he had isolated the source of maleness.

But two years later, at the end of 1989, another group reported studying four XX males whose bits of translocated Y failed to include the magic sequence. "Back to the drawing board," said Page, suggesting that his earlier finding had probably isolated a male-determining gene all right, but perhaps not the primary one. In any event, he had reduced the search to a very small region; only 0.2% of the tiny Y-chromosome needed further investigation.

Now it's all over but the shouting, says the July 19, 1990 issue of *Nature*, which tells us precisely which tiny piece of DNA on the human Y-chromosome probably does the trick. (The only other trick this stunted Y-chromosome is known to perform is the production of excessively hairy ear rims!) With the exact DNA sequence now pretty well identified, scientists can turn their attention to understanding which protein it encodes. This substance is literally the stuff that little boys are made of.

Barr body, Barr body, who's got the Barr body?

When it was determined in 1949 that normal mammalian females always show Barr bodies and normal mammalian males don't, an obvious next step was to test those with abnormal sexual characteristics to investigate their Barr body situation. Klinefelter males came immediately to mind, as did females suffering from Turner syndrome.

This complex of symptoms was first described in 1938 (by a Dr. Turner, of course), four years before the first Klinefelter paper appeared. Turner women all exhibit sterility, a "webbed" neck, and deformity of the elbow; instead of having extraordinarily long legs like their male counterparts, those with Turner syndrome are all exceptionally short; unlike the males, they're not mentally retarded; and instead of being fairly common (1 in every 500 live male births), the females are very rare (1 in every 3500 live female births. This is because most Turner fetuses are spontaneously aborted; Nature apparently makes no related attempt to select against the birth of those with Klinefelter syndrome.)

Despite their marked differences, it was clear that these similarly sterile men and women were somehow sexual reciprocals of one another, although no one had a clue what caused their curious conditions. Thus there was great excitement when it was discovered in the early 1950s that the Turner women often lacked the typically female-determining Barr bodies, whereas the Klinefelter men all exhibited them! And yet there was no question about the sex of either camp. What was going on with their seemingly upside-down genetics?

By 1956, the correct human chromosome count of 46 was finally determined, and decent karyotypes could at last be made. By 1959, it was possible to see that most Klinefelter men had an extra X (had 47 chromosomes and were XXY) and that the Turner women were missing an X (had 45 chromosomes and were X0). Once the Lyon Hypothesis was accepted, things didn't seem so mysterious. Of course most Turner women didn't show any Barr bodies — they didn't have any extra Xs to be inactivated. And of course the Klinefelter men did show Barr bodies — they all had extra Xs to be inactivated, just like normal females.

In fact, 20% of the Klinefelter men had quite a lot of extra Xs. Rather than being just XXY, they were XXXY or XXXXY or even XXXXXY, and some were complicated mosaics with extra Xs in several different cell lines. When their cells were studied, they showed not just 1 Barr body, but 2 or 3 or 4 — always one less than the total number of Xs involved. About one-third of the Turner women were mosaics too, mostly X0/XX; such mosaics showed Barr bodies in their XX cells, the ones that were just like those of normal females.

It seemed that some mechanism was at work that said, in effect, if you're a mammal, male or female, one working X-chromosome is all you're allowed; any extra ones will simply be inactivated. How did this curious state of affairs arise? What was it all about?

Inequality of the sexes

The Klinefelter men and the Turner women (and the calico cats) were leading me at last toward the answers to some questions posed in innocence and ignorance so long ago. To start with, do females, with their two Xs, have twice as much genetic information about certain traits as males? If so, isn't this an unfair advantage?

Reviewing these questions in the original, I see that they were scribbled in great haste on a pad of yellow paper in the spring of 1988. In those days my mind was like the hot mud pots of Yellowstone. Thoughts kept bubbling to the surface and breaking with a sudden splat in the form of questions. I was clearly hurrying to record these questions quickly, to capture them before they evaporated, and it never occurred to me then that their exact wording might be troublesome or significant.

When I wrote the phrase "twice as much genetic information," I was thinking vaguely about how each of the genes was responsible for producing a particular protein whose job it was to help specify a certain trait. Because females had twice as many Xs as males, they'd have twice as many X-linked genes, and these would produce twice as much gene product. This didn't seem fair — it seemed unbalanced.

To answer these questions in reverse order, and in the way that I had intended them at the time: Yes, it would be unfair; and No, nature won't allow it. Inactivation of supernumerary Xs is one of her several methods of what's called "dosage compensation" — the regulation of the amount of gene product that's manufactured, no matter what the chromosome situation is like.

In fruit flies, where the males are normally XY and the females XX, the males make up for their missing chromosome by working twice as hard: A male X produces twice the gene product of a female X, but because the female has two of them, it comes out even. In mammals, where the males are also XY and the females XX, the opposite approach is used: To compensate for the imbalance, extra Xs are inactivated. Some slight imbalance in protein production remains, however, because (1) a tiny number of genes are active on the Y that aren't represented on the X and (2) not all genes on the X are actually inactivated (at least in humans).

Evidence for partial inactivation comes from the fact that Turner females and Klinefelter males are both abnormal, the females being much worse off than the males. If the extra X of a normal female were completely inactivated, then she would have only one X in good working order — she would, in effect, be X0, just like the Turner women. Yet the Turner women are far from normal: They're invariably sterile (from ovaries that don't work or an underdeveloped uterus) and have many other problems besides. Similarly, if all the extra Xs of Klinefelter males were completely inactivated, why wouldn't they be just like normal XY males? Why should their symptoms become more pronounced as the number of extra Xs increases?

The answer is that apparently a few genes at the tip of the X-chromosome are spared from inactivation. The extra presence of these genes in men seems to produce tallness, sterility, and mental retardation, and their absence in women seems to produce shortness, sterility, and a webbed neck. This is clearly a complicated mechanism, and no one really understands how it works.

Let's revisit that first question again: Do females, with two Xs, have twice as much genetic information about certain traits as males? Using a more literal interpretation than before, we find that the answer comes out yes instead of no. Of course normal female mammals have twice as much genetic information as males, at least where the X-chromosome is concerned: They've got one type of X inherited from their mother and another (possibly very different) type of X inherited from their father; the males, by contrast, have no choice. One type of X, inherited from the mother, is all they get.

Thus, although the quantity of X-linked genes (and hence the total gene product) is roughly equalized by X-inactivation, the "quality" is better in females because they have the opportunity for much greater genetic diversity. Bad news from an X-linked gene inherited from one parent may be counteracted by good news from the matching X-linked gene of the other. By contrast, if a male inherits a bad-news gene from his mother's X, it's just bad news — his father has contributed only a shrimpy Y and has nothing further to say about the matter.

But what genes do people have on the X that aren't on the Y? Does this account for baldness, color blindness, muscular dystrophy, and hemophilia, which usually afflict males only? Yes, but these are only a few of the approximately 120 abnormal traits that have been definitely linked to the X-chromosome; there's also deafness, mental deficiency, spastic paraplegia, Parkinsonism, cataracts, night blindness, and nystagmus (the rapid and uncontrollable oscillation of the eyeballs that Doncaster was studying in 1913). Many more are suspect and are under investigation.

Much early knowledge about the genes on the X was acquired through pedigree analysis, some of it undertaken long before the mechanisms of inheritance were understood. Passages in the Talmud, for example, indicate that it was known in ancient times that color blindness is usually transmitted only from mother to son; the famous pedigree of Queen Victoria shows many unfortunate male descendants, but no female ones, afflicted with hemophilia, "the disease of kings." (Note that these men couldn't pass hemophilia, or any other X-linked trait, along to their sons, because their contribution to male offspring was necessarily a Y, not an X.)

Occasionally, of course, "random" X-inactivation results in a vastly disproportionate amount of maternal or paternal Xs being done away with. In these rare cases, a female may be afflicted with one of these traditionally male conditions if she's left with too few of the other parent's genes for counterbalance. A female will also, of course, be afflicted with such a condition if she has the unusual bad luck to inherit two genes specifying it, one from each parent.

Many women are carriers of genes that don't affect them but may seriously afflict their offspring. Consider, for example, a woman who shows no signs of color blindness (she can argue at length about the differences between mauve and puce) but who has actually inherited a gene for color blindness from one of her parents. If she mates with a man with normal vision, half her sons will be color-blind but none of her daughters, although half of them will be carriers like herself. If she mates with a color-blind man, half her sons will still be color-blind as before, but now half her daughters will be color-blind as well, and the other half will be carriers. (If this description seems counterintuitive to you, as a lot of the mathematics of genetics does to me, make yourself a Punnett Square to see what's actually happening.)

So it goes in the still unbalanced, still unfair XX/XY world, despite the equalizing effects of X-inactivation. Because of their greater genetic diversity, women are by no means the weaker sex; they continue to wreak havoc, on men for the most part, while remaining largely unscathed in the process.

Perhaps further evolutionary developments are under way that will equalize the sexes in some distant future. Consider the fact, for example, that slightly more human males are born than females, all the time and in all places. (The differential for white Americans is 6 %.) This may be because the Y-bearing sperm are somewhat lighter than their X-bearing competitors, swim faster, and thus have a higher probability of being the first to hit the target; or it may be that they have higher viability and are hence more numerous. Any of these possible factors may be part of some complex compensatory scheme to address yet another result of XX/XY inequality: Currently the life expectancy of the human female is always greater than that of the human male, at all stages of development.

Hypertrichosis of the pinna (hairy ear rims!)

Well, of course I couldn't resist tracking down the hairy ear rims (a condition known properly as hypertrichosis of the pinna). Even though they don't have any observable connection to calico cats, they provide an interesting example of sex-linked inheritance. In this case the linkage is to the Y instead of to the X, so rather than mothers afflicting half their sons and causing half their daughters to become carriers, this time fathers pass the trait along, afflicting all their sons exclusively.

I first encountered hairy ear rims in pictures of three Muslim brothers from southern India. They were staring at me rather desperately from the pages of the freshman biology text *Human Genetics*; all seemed to have stiff mustaches growing out sideways along the edges of their ears. (I learned later that the first pedigree of hairy ear rims was published in Italy in 1907 and that Mediterranean Italians, Australian aborigines, Nigerians, and Japanese have also been known to sport such appendages. But apparently genes on other chromosomes than the Y are sometimes thought to be responsible.)

Several papers about Y-linked inheritance have been written by an Indian geneticist named Dronamraju. In the late 1950s, when working in Calcutta, he was visited by an American colleague who noticed his hairy ear rims and suggested that he investigate his family pedigree for other examples. But Dronamraju did even more. Because it was then an eleven-day journey by bus and train from Calcutta to his ancestral state of Andhra Pradesh, he spent the time contemplating the ears of other passengers along the way. As he says,

> By changing my seat more than once I was able to make an approximately complete inventory of a compartment, no one being scored from a distance of more than 1.5 metres.

(One can't help wondering what his fellow travellers must have thought of his restless behavior.) He scrutinized women as well as men but found no examples (although he was sometimes thwarted by the burkhas the Muslims wore to veil their faces). Nor could he find any informant who had ever seen or heard of a woman with this characteristic. He points out that

> This particular hairiness of the ears is noticed and remembered because of a certain belief connected with it in the society of which people in the pedigree formed a part. It is considered as correlated with longevity of the man having it. For a man this is regarded as extremely fortunate. Contrarily any excess hair on a woman is regarded as unfortunate, as this again portends longevity and therefore the danger of outliving her husband.

In the course of his familial quest, he also tried to examine old photographs but was frustrated by the fact that the edge of the ears was often blurred and out of focus. And in more recent times "two photographers have told me that it is their practice ... to obliterate what modern people may consider as a blemish."

Despite these difficulties and the general squishiness of his family data (some provided from memory about people who were no longer living), Dronamraju grinds out a lot of statistics and even two alternative hypotheses explaining how hairy ear rims might be inherited. He concludes, however, that the trait is indeed Y-linked, that "the age of onset seems to be in late puberty," that "the density of growth and the length of individual hairs both increase with age," and that "about 6% of an unselected sample of 345 adult males in Andhra Pradesh were found to show such hypertrichosis."

A few years earlier, in 1957, Curt Stern wrote a paper entitled "The Problem of Complete Y-Linkage in Man." He also discusses pedigrees, some very old, purporting to show Y-linkage for peroneal atrophy, bilateral radio-ulnar synostosis (the fusion of adjacent bones near the elbow), camptodactyly (bent little finger), cataracts, adherent tongue, foot ulcers, keratoma dissipatum (skin defect), webbed toes, ichthyosis hystrix ("porcupine men"), color vision, and abnormality of the external ear (not having to do with hair). And he discusses hypertrichosis of the pinna and concludes that it's the only characteristic for which Y-linkage holds up under scrutiny.

Perhaps more genes will be found there in the future, but for now the human Y-chromosome is famous only for making all its products male and for giving a small percentage of them very hairy ear rims.

Slouching toward the Promised Land

Even before we installed our fancy new electronic cat door, I realized that freedom of access for the cats, like many other kinds of freedom, could have both good and bad aspects. A good aspect, I imagined, would be our release from bondage: enough of this leaping up to open the laundry room door whenever we heard Max's faint and plaintive meow or George's vigorous and forbidden scratching on the various glass doors that bear the imprint of his muddy paws. A bad aspect, I imagined, would be the loss of the daily pleasure of finding them waiting for us on the covered porch, eager to see us arising at last to let them in for breakfast.

I was wrong on both counts. Even though George and Max are now able to come and go at will, they still scratch and meow to be admitted if they know that we're about; not wanting to seem unfriendly, we still find ourselves at their beck and call. But, as if in compensation, they also still give their morning greeting on the porch, clustering around our legs and acting as though they can't get in (or along) without us.

We hadn't envisioned the main bad aspect, however, which has to do with tender offerings. Initially such tributes, usually in the form of little piles of intestines, were placed on a certain stepping stone in the middle of the lawn. Then, as my bare feet discovered to my horror one morning, the favored place became the rough mat outside the laundry room door. Now the catflap has allowed the mat *inside* the door to become the preferred altar, and the latest sacrificial victim appears to have been a very large rodent indeed.

Having infiltrated (and anointed) the laundry room, Max is determined to expand his territory. If the heavy sliding door that separates the laundry room from the kitchen is not completely closed, Max's black and white paw will soon be seen wiggling through the tiny opening. Then slowly but persistently, and with considerable effort, he succeeds in making this crack wider and wider. Max understands the Maginot line dividing the linoleum of the laundry room from the hardwood of the rest of the house and toes this line as long as anyone is watching. But he is ever hopeful, ever on the lookout for his chance to invade.

Recently I was awakened by his enquiring voice, which seemed surprisingly close by. And there he was, up in the loft, just outside our bedroom door, very eager to join us for a little midnight visit. (Our bed no doubt has been his main objective all along.) Investigations downstairs revealed a six-inch opening in the sliding door. Wondering which of us had been the negligent party, I put Max in his basket in the laundry room, spoke to him sharply, and slid the door shut very firmly behind me.

In the morning the six-inch crack had reappeared, but Max was virtuously curled in his basket, watching me with enormous and knowing eyes. Shortly thereafter my bare feet encountered his latest offering, which, as usual, was laid carefully in the center of the mat — this time the mat in front of the kitchen sink.

I know he'll win eventually (and so does he), but first we'll try to rig some sort of contraption on that sliding door. It will no doubt prove to be a weak defense against so determined a dark invader.

How many chromosomes to a duck-billed platypus?

When I first learned that people have 23 pairs of chromosomes and cats 19, that somehow seemed about right. I imagined that the great apes would have slightly fewer than humans, dogs would have about the same number as cats, mice would have fewer, and peas fewer still. It seemed likely that people would have the most. But, like many of the things I imagined about genetics, it didn't turn out that way at all. Chickens, dogs, horses, goldfish, plums, and the duck-billed platypus all have more chromosomes than people, and mice have more than cats.

Does it matter? Not much. It isn't the chromosome count that's significant, but the number and type of genes an organism has to work with. And these are happy to distribute themselves on large chromosomes or small, over a few or many. Of course if there were too few, perhaps only a single pair, then there would be much less genetic diversity because the genes would be linked by all living together on a single block. But as long as there are enough blocks, and there's matching real estate across the street for a partner to occupy, a gene can go about its business of producing a particular protein without much regard for what part of town it's been assigned to live in.

There were other surprises as well. I discovered that for all mammals, from shrews to chimpanzees, the total amount of DNA at work in the nucleus of each cell with a normal chromosome complement is almost exactly the same — about 7×10^{-9} milligrams. Even more surprising, the X-chromosome in all mammals always contains about 5% of this total. (This figure was arrived at in a wonderfully simple and straightforward manner. Karyotypes of various mammals were made and enlarged to 6300 times their normal size. Images of all the chromosomes were projected onto white typing paper, traced with a sharp pencil, and then carefully cut out with sharp scissors and weighed on a precision balance. By comparing the weight of the X-chromosome with that of the complete chromosome set for each mammal, investigators arrived at the more or less constant value of 5%.) This constant value supports other evidence that the genes on the X haven't changed much for a very long time. An ancient X seems somehow to have been preserved more or less in its entirety, and whatever genes live on the X of one mammal probably have counterparts on the X of another. For example, X-linked hemophilia has been found in dogs and horses as well as in humans; X-linked anemia also occurs in both mice and men.

Such similarities are not, of course, found with the other chromosomes because it's hard to tell what their analogs across species might be. In the course of evolutionary development, old chromosomes have split in two to produce new ones, some have fused and been reunited, and parts of chromosomes have been translocated onto others in a constant reshuffling of gene locations. That's how the wide variety in size and number of chromosomes came about. The stability of the X, by contrast, is due to its isolation. It doesn't mix it up very much with the other chromosomes or with its miniscule partner the Y, with which it has virtually no gene locations in common.

The case of the shrinking Y

What did nature have in mind when she left so little space on the Y? Why are the sex chromosomes the only pair to be different in size and shape from one another? Isn't this bizarre? Before answering these questions, I must first apologize for their naive wording. Nature, I now know from reading Richard Dawkins, has nothing in mind; things just happen, which cause other things to happen, and those mechanisms that result in successful organisms are preserved. And the sex chromosomes aren't necessarily an unmatched pair: They're the same size as each other in female mammals, in male birds, and in both sexes of various other species. What I meant to ask was why the Y is so different in size and shape from the X. Isn't this bizarre? If this discrepancy occurs in the sex chromosomes, why isn't it seen in some of the other chromosome pairs as well?

Yet a different kind of question arose concerning the freshman biology text that first brought this curious fact about the X and the Y to my attention. The prose of this book was both nonchalant and noncommittal:

> Of the 23 pairs, 22 are for all practical purposes perfectly matched in both sexes and are called autosomes. The remaining pair are called sex chromosomes, and though the members of this pair are apparently identical in women, they are not identical in men.

Period. Why didn't *it* say, "How bizarre!"? Like other texts I consulted later, it made no attempt to remark on the strangeness of this situation or to explain why or how it might have come about. I had a similar reaction to descriptions of X-inactivation, a process that also struck me as remarkable. A biology text different from the one just quoted devotes less than a third of a page (out of 1159!) to this fascinating process and treats it very dully (even though it uses a great picture of a calico cat as an illustration).

The book that had been sequestered by the mad librarian, however, attempted to provide answers to these questions. It was entitled *Sex Chromosomes and Sex-Linked Genes* and written by Susumu Ohno in 1967. As I settled in to peruse it for the second time (in a photocopy obtained from a medical school nearby), I was suddenly grateful to that maddening and officious young man who had hidden it from my then unworthy view. Several years had passed since my initial brush with this material, and I was surprised and delighted to find how much it had improved. It had become nearly comprehensible!

Struggling though its many pages, however, and its many theories about evolutionary development, I came to appreciate why the texts had punted — an entire book was necessary to provide serious answers to these questions. I still couldn't follow all of Ohno's complex arguments, some of which seemed to elude me just as I thought I was getting the hang of them, and the whole subject is still virgin territory, as Brian Charlesworth (who doesn't even reference Ohno) indicates in the March 1, 1991 issue of *Science*.

The important central ideas, however, seem to be that the X and the Y, like all the other chromosome pairs, used to be identical to one another in size and shape. In some organisms this is still true. The snakes in particular provide an interesting picture of evolution at work. They fall into three different classes with regard to the X and the Y (or more accurately the Z and the W; for snakes, as for birds, the females are ZW and the males ZZ). In boa constrictors, the Z and the W are still a matching pair; in gopher snakes, they're about the same size but the position of the centromere has changed in the W, making them look unbalanced; in poisonous snakes like the side-winder rattler, the W is tiny, as it is in birds (and as the Y is in most mammals).

Why and how this came about has to do, of course, with sex. The first signs of life appeared over three billion years ago. For about the first billion years, ameba-like creatures happily and exactly replicated themselves, ignorant of the joys and benefits of sex: They just used mitosis to make copies of their chromosomes and split. Then the new, more complicated process of meiosis evolved from mitosis, and sex cells were formed. At first these sex cells all looked alike, but over the next billion or so years they differentiated into large, relatively sedentary eggs and tiny, very mobile sperm. (Nature may be mindless, but she certainly has patience!)

Although meiosis was a complex and dangerous process fraught with possibilities for error, it was here to stay because of the obvious advantages it provided. It created genetic diversity through recombination and thus enabled species to adapt to new environments, and some of its errors produced mutations that led to the development of new and wonderful things. Ohno theorizes that initially there may have been a pair of genes that specified either one sex or the other. Then other genes having to do with further aspects of sex discrimination came to collect on the same chromosomes, through the process of gene duplication (another sort of error in which a gene doubles itself, increasing its size and that of its chromosome, and allowing for further functional development). Once these sex chromosomes were established, they had to be inhibited from crossing over: If they were allowed to mix it up, then male and female characteristics would come to occupy the same chromosome, and chaos in the form of a largely hermaphroditic species would result.

It's clear that this isolation of the sex chromosomes has occurred in many species and has resulted in the preservation of an ancient X and Z. What's not so clear is the story about the degeneration of the Y and W. Degeneration is certainly the perfect word for describing the loss of genes; apparently almost all of them on the Y and W were lost except the sex-determining ones. Once this started to happen, however, some method of dosage compensation was clearly needed. Perhaps at first it was a *Drosophila*-like method, the genes on the X having to work twice as hard to make up for the lost genes on the Y. This worked fine for the males, which had only one X, but because both female Xs worked twice as hard too, they now produced a lot more gene product than was needed or desirable. This eventually resulted in X-inactivation so that things could be evened out once more. Ohno thinks this must have happened about 100 million years ago.

As a result of complete and redundant replication of their entire Xs, a few species of rodents have extra large X-chromosomes. Instead of having the normal 5% of the DNA, some, which resulted from a doubling of the original X, have 10%; others, which arose through tripling, have 15%. To even things out for them, X-inactivation not only incapacitates one of the female Xs but also incapacitates half or two-thirds of the single male X, depending on whether a doubled or a tripled X is involved. This extra wrinkle on X-inactivation makes it seem even more remarkable than before.

For some reason, no such dosage compensation has developed in the birds. They continue to lead a highly unbalanced ZZ/ZW life, the females taking the brunt of any differences instead of the males. Ohno remarks that "this failure is one reason why birds have not escaped the status of feathered reptiles." I'm not entirely sure what he means, but it certainly sounds like a bad fate to me, and it may be related to the fact that birds have only about one-third the DNA of mammals.

Those who want a fuller explanation about the shortness of the Y and W are invited to explore Ohno (who provides many further examples about animals) or Charlesworth (who argues on the basis of plants), or the copious references that they both supply. Perhaps we'll never know what caused the Y and W to shrink or how the various methods of dosage compensation emerged, but it's important to wonder about these things, to think them remarkable, and to try to understand how these curious facts fit into the constantly but very slowly changing picture of evolutionary development.

George gives blood

Although I'd been able to curb my curiosity about current Alzheimer research, my curiosity about George's genetics finally got the better of me. It had been a long time since I'd struggled to understand and describe the fairly simple mechanisms of producing an XXY George by non-disjunction, but I still wasn't any closer to knowing whether he was actually one of those or a more complicated kind of cat; I had no idea what method George's parents — or maybe George himself — had used in his construction. George might have any of the abnormal chromosomal configurations shown in Figure 8 or perhaps even one that hasn't been encountered yet.

A doctor friend said that chromosome testing was easy, at least for the cat. All that was needed was a little scraping from the inside of the cheek. George would hardly notice. This buccal mucosa could be used to produce a karyotype in which all his chromosomes would be arranged in neat rows for inspection. From this the truth would out. Was he sex-reversed XX, a simple Klinefelter XXY, or a complicated, possibly novel, chimera or sexual mosaic?

The large medical school in the next county does chromosome testing all the time, working usually with amniotic fluid in the search for sickle-cell anemia, Down syndrome, and other chromosomal abnormalities (like XXY and XYY) that bode ill for fetuses and their parents. Maybe, I thought, I could find a technician willing to moonlight. And then I imagined the scene in the lab: "Hey Joe, look at this! Only 19 pairs! Did ya ever see one of these before? What'll we tell 'em? They'll have kittens!"

These fantasies over, I contacted the nearest vet school instead and asked if they could do the job. Sure, they said, delighted to be of service. But buccal mucosa was out of style; they wanted George's blood. It could be delivered either inside the cat or inside a test tube, whichever was more convenient. If a test tube was used, however, it needed to be delivered quickly so that the sample would be nice and fresh.

The tube was really the only choice, because George hates cars and is glad to let you know about it. Neither he nor I would have been able to stand the eight-hour round trip involved. But how to get the blood out of the cat? A lengthy journey to the vet would still be necessary and George wouldn't like that either. But I knew he'd tolerate it better if my granddaughter Valerie were around, so I scheduled the blood-letting for her Easter holiday.

Valerie is the perfect assistant. At fourteen, she's already certain that she wants to be a vet and works after school at an animal clinic. (Her boss has trouble believing in Valerie's description of George — "Are you *sure* it's a male?" — and has never seen one before either.) Although George is very choosey about his human friends and doesn't usually tolerate much touching, he always melts in Valerie's skillful hands. She cradles him on his back and lazily scratches his tummy; his eyes glaze as he goes limp and drifts off into some cat Nirvana.

Valerie soothes George in this manner, making his trip to the vet tolerable, and then professionally accompanies him into the inner sanctum to help while blood is drawn from his jugular. (I'm just as glad to be restricted to the waiting room where I don't have to watch and can feel guilty all by myself. Is this trip really necessary? Is George suffering? Do I really need to know what his chromosomes look like?)

Two tubes are filled, each containing ten milliliters of George's deep red blood. A small amount of heparin is added to prevent clotting, and then the tubes are gently inverted several times and placed carefully in a thick styrofoam container. With our precious cargo on board, we rush home, leave George in Max's tender care, and embark at last on our urgent odyssey; it's already nearly ten and we've been warned not to stop for lunch.

Valerie and I are treated like visiting royalty when we arrive around two at the Serology Lab. We've brought them something different: not horse blood, which is their daily fare, nor llama blood, a more recent interest, but cat blood, which is now seldom tested. I'd imagined a modern tower of a building filled with gleaming stainless steel equipment and technicians in starched white lab coats. Instead we find ourselves in a trailer where friendly people in casual clothes are peering into microscopes or bending over trays of gel from which electrical wires protrude.

Our guide explains that their testing of horse blood has to do with the verification of pedigrees. The work is very routine, but about four times a year there's considerable excitement in the lab: Twice a year some of the old electrophoretic equipment they use to determine blood types catches fire, and twice a year their work unearths some evidence of fraud in the horse-trading world, which generates considerable heat as well.

We spend an hour or so touring the lab, trying to understand the path that George's blood will follow during the next few weeks of analysis. But we're too tired and it's all too confusing; we retreat home to be with George and to read about it in the literature.

The making of a karyotype

Early pictures of chromosomes, like those made by Walter Sutton of the great lubber grasshopper in the early 1900s, were probably made by cutting up some tissue with a sharp knife, placing it in some fluid that would kill and fix it rapidly, and then mounting the resulting slivers inside blocks of paraffin or other similar media. These in turn were cut into very thin sections, mounted on slides, and stained so that the chromosomes could be seen under a microscope. Pictures were made by placing a piece of drawing paper below the microscope's stage and tracing what was seen projected on it. This worked fairly well as long as the chromosomes were large and there weren't too many of them. But the fact that a correct human chromosome count wasn't obtained until 1956 shows clearly that this technique had serious limitations.

The first reports of the chromosome count of *Felis domestica* were published in 1920. There were two of them that year, written by Germans who came to different conclusions. The first declared that cats had 35 chromosomes and produced two different kinds of sperm, one with 17 chromosomes on board and the other with 18; the second declared that cats had 38 chromosomes, with 19 showing up in every sex cell. In 1928, and again in 1934, the Japanese geneticist Minouchi published papers confirming 38 as the correct number and providing pictures to prove it.

Minouchi's description of his research methods is a bit harrowing. He explains that he got his hands on some male cats that were attacking an actinida at the Institute of Dendrology at Kyoto Imperial University. (Actinida is a woody vine with edible fruit that, like catnip, is very attractive to cats.) These hapless cats were caught biting the bark and mewing while they danced about it. For these transgressions they were "killed by decapitation and the testes taken out from the body immediately. At the same moment they were cut into small pieces and dropped" into a fixative solution and then sectioned, viewed, and drawn as described above.

More modern (and humane) techniques involve the use of tissue culture in which cells are grown in some synthetic medium. Although buccal mucosa or skin biopsies from various parts of the body were commonly used in the past, the preferred substance is now whole blood. Because contamination from molds and bacteria is a serious problem, great pains are taken to keep everything sterile, and fungicides and antibiotics are routinely used. To ensure that there will be an abundance of cells in the proper state for viewing, a substance that induces mitosis, such as pokeweed or an extract of navy beans, is also added. This peculiar brew of blood, synthetic culture medium, anti-contaminants, and natural mitogens is then placed in an incubator and kept at the body temperature of the organism being studied (in George's case, at 38.6 degrees Celsius) for several days.

Harvesting involves putting the culture in a centrifuge and spinning it hard, a number of times, to obtain the white blood cells that will be left at the bottom of the tube. A few hours before the first centrifuging is done, a substance is added that inhibits the formation of spindle fibers. (Colchicine, made from the poisonous root of the autumn crocus, was commonly used in the past but has now been supplanted by its less romantic-sounding but more effective synthetic analog colcemid.) With no spindle fibers around to pull the chromosomes to the opposite ends of the cell, most cells are arrested in the lining-up phase of mitosis; thereafter, no cells are found in the poling-up or splitting-up phases.

After this initial spin, potassium chloride is added; it causes the white cells to swell up nearly to the bursting point. This separates the chromosomes from one another and makes them easier to view. A fixative is also added to toughen the swollen cell membranes so they won't burst when they get whirled around some more. The sample is then put back into the centrifuge for another ten minutes at 1200 rpm. This happens to it several more times; in between, accumulated debris is discarded and further additions of fixative are made.

Eventually the white cells remaining at the bottom of the tube are dropped onto very clean slides and left to dry. They can then be treated with an enzyme called trypsin and stained with various materials so that the distinctive bands that characterize each chromosome can be seen (Giemsa stain for so-called G-banding is most commonly used.) These bands help to identify the members of each pair, which are sometimes hard to distinguish if the organism has many chromosomes of the same size and shape.

Once a good slide is obtained, a photograph of it is taken through the microscope and projected for enlargement. (In George's case, we were told, the lens of the microscope will magnify the chromosomes 63 times, and the lens of the camera will enlarge them another 10 times; thus the chromosomes in the resulting negative will be 630 times normal. This negative will then be enlarged until the chromosomes are approximately 5000 times their actual size.) Each enlarged chromosome is then carefully cut out of the positive picture with sharp scissors and placed beside its counterpart on some light-colored cardboard. The chromosome pairs are arranged according to the San Juan convention, from longest to shortest, and grouped together according to the placement of their centromeres, with the sex chromosomes always coming last (as seen in Figure 6). This arrangement is then photographed again to produce, at last, the finished karyotype.

Valerie and I (and George) have done our part and now we just have to wait and hope that the culture will grow successfully, which it sometimes doesn't. Even if it does, it may be months before we learn the secret of George's genesis.

The Poon Hill tower

Friends came bearing slides of their recent adventure in Nepal. They'd completed the same 250-mile circuit of the Annapurnas that we'd trekked in 1980, and we experienced a thrill of nostalgic recognition at seeing certain mountains — and even certain Nepali people — that we remembered from before. But we were dismayed to see power poles and guest lodges, and even a few rough roads that could be negotiated by truck, none of which had been there a decade earlier. The thing we found most upsetting, however, was not a presence but an absence: The wonderful rickety old tower on top of Poon Hill was there no more.

I'd studied its picture often in our photo album, hoping to build a replica here on the namesake of its setting during the construction of our cottage. It was easy to imagine, however, what the county building inspectors would think of *that* project, and we never got around to asking them. Instead we bought an ugly steel water tank (practical and necessary for fire protection), painted it disappearing brown, and built a sturdy deck on top of it from which to admire the blue Pacific.

When the nights are dry, as they often are in this fifth year of serious drought, we sleep fifteen feet in the air on top of our inferior tower. The only ambient light is from the moon, so the stars are often very bright — and sometimes falling. We watch the satellites trace their rigid orbits in the sky and wonder what new folk legends are being created in the depths of a New Guinea jungle about The Stars That Move and how they came to be that way.

George and Max sleep on top of the tower too — actually, they sleep on top of us. As we climb the steep stairs toward our outdoor loft, we often find Max waiting, comfortably in possession on the sleeping bag, wondering what has kept us awake so long. He and George will go off foraging after midnight, but in the morning they will both be there, giving each other a bath at our feet or playing happily with their latest acquisition: a small bird, mouse, or mole. So far, I'm happy to report, no snakes have been brought as tribute, but our sleeping bag is showing many signs of carnage.

The tower, with its narrow winding staircase, is both a place of refuge and a vantage point from which our felines can survey their territory. From their high perch on its railing they scan the meadow, calmly watching, knowing that the overhanging limbs of the pines provide an easy escape route should an enemy, canine or feline, have the temerity to attempt the stairs. So they like our Poon Hill tower, not knowing how poorly it compares with the original we are now mourning.

Max knows

George and Max have come to know our habits well. On cold mornings, when they're still lively and covered with burrs from their night's adventures and I'm still dazed and dreamy from my night's repose, I can be counted on to walk down through the orchard to the Poonery to light the heater in preparation for the morning's work. After breakfast, George usually follows me down again, to curl up on the couch in this now warm sanctuary; there he'll laze away the useless daytime hours while I read and write and try to be productive.

Max still prefers the laundry room, despite Houdini's desecrations, and seldom comes to join us. It's clear that writing doesn't interest him — but construction, that's another matter. When such projects are in progress, Max is always underfoot, eager to be of assistance. He makes an excellent supervisor but, to our sorrow and his, is sometimes banished to the garage to prevent his being painted, squashed, tangled in barbed wire, or otherwise damaged.

The garage is Max's province in other ways as well. Occasionally he asks to be admitted to check on the rodent population; at other times he watches closely as we approach the car to find out what we're up to. If it's the weekly grocery run, that's fine with him; if we're taking only trivial items (windbreaks, binoculars, and towels for the beach or garbage cans for the local dump), he pays little attention. If, however, he sees us gathering the tent and the backpacks and the sleeping bags — those harbingers of travel and desertion — then he dogs our footsteps and complains bitterly, or leaps through the window and sits in the front seat of the car, or lies down behind its rear wheels, indicating emphatically that we're not to go away and leave him. Once he even smuggled himself on board a van that was taking us to the airport. When the van braked sharply to avoid an approaching horse trailer his startled meow revealed his presence, and we were forced to turn around and deposit him gently but firmly in the meadow.

But wherever we go, we can always count on his enthusiastic welcome when we return. If it's dark, our headlights usually pick up Max's white bib, bobbing up and down as he races from the woods toward the driveway. There he'll roll and roll, wriggling his back in the dusty gravel of the drive, making his white stomach maximally available for scratching and stroking. If it's light, Max is usually watching from the laundry room window or the covered porch. Again, he'll leap into action, racing for the gravel of the drive. Sometimes, if he's late, he'll run right past us as we arrive, dashing down the stairs as we come up them. This forces us to return to the driveway where the rolling ritual is always held.

In light or dark, George will come to greet us too, but much more indirectly. Anticipating our entrance, he'll bound up the stairs in front of us, with his long hind legs hopping in unison like a rabbit's. Then he'll wait impatiently at the top to be admitted, while we trudge more slowly, hauling heavy bags of groceries or other supplies. The cat door allows him to come and go whenever he likes, but he hopes to please us with this little show of false dependence.

Max gets a lot of attention for his rolling, so George has tried it out a time or two himself. But his performance is studied, copied, and unspontaneous — it's clumsy and seems unnatural. One morning he watched as we descended sleepily from a night spent on the Poon Hill Tower. We could see him calculating the effect before he rolled in an awkward fashion on the cement footing at the bottom of the stairs. Once he surprised us by rolling in the freshly turned earth of the vegetable garden. And this morning, when I came down to light the heater, he rolled a little on the rug in the Poonery. Progress, I think, with even some independent variations on the theme (Max is into gravel only). Perhaps George will become less aloof, more touchable, more accessible, more like Max. And then I realize that I like George just the way he is: difficult, independent, and detached.

9

The Late Calico Papers

XXY, that's why

Waiting for results from the Serology Lab makes me restless and impatient. To release my nervous energies, I delve deep into the stacks again, searching for the next threads in the story about George's predecessors and their contributions to the history of genetics.

During the ten-year period between the discovery of Barr bodies (in 1949) and the announcement that Klinefelters are XXY (in 1959), there are only a few desultory attempts to work further on the problem of the male calicos. In 1956, the Japanese researchers Komai and Ishihara (about whom we'll hear more presently) dust off some old theories of Doncaster's and suggest that these very rare animals are the result of very rare crossing over between the X and Y chromosomes. Two other papers also appear in 1956. In one, a pair of Americans try to revive Doncaster's freemartin theory of 1920 and suggest again that fused placenta of fetuses of different sexes may be the answer. In the other, Ishihara studies the testes of six male calicos to see how well they're functioning.

He finds that four of the six show no ability to manufacture sperm, but two of them are just as good at it as those of any normal tom. He perceives no differencess in chromosomal constitution among the six:

He declares that their cells all show the normal complement of 19 pairs and that they all have standard male XY chromosomes. (Like the researcher from the University of Edinburgh described earlier, he probably still can't really see what he's doing.)

In 1957, another one-pager appears in *Nature* announcing the existence of an extremely fertile tortie tomcat named Blue Boy. His name reflects the fact that he has "blue," or dilute, pigmentation (*d*) instead of dense pigmentation (*D*) and hence has the genes *dd*; he also has short, curly rex hairs (*r*) instead of a normal coat (*R*) and hence has the genes *rr*. His father was the first English rex cat recorded; his mother was a tortoiseshell known to carry the genes *d* and *r*.

Because of his reliable pedigree, Blue Boy is much more interesting than earlier animals whose parentage was questionable. He's also interesting because he's the first male calico known to produce more male calicos — and not just one or two, but eleven out of the forty-three kittens he'd sired by 1957. (Only three had reached maturity by then and they were all sterile.) As is the breeders' typical practice, Blue Boy has been mated only to other calicos — six of them, all relatives of his — and all possible variations on the color schemes, in both sexes, have occurred.

This is rich and wonderful material, but the authors don't really know what to make of it. They proceed to review other old theories about partial this and that and propose that "in Blue Boy the gene for yellow has become partially sex-linked, instead of being completely so." It's not clear how Blue Boy goes about the business of self-replication, a trick which all other fertile male calicos have so far failed to learn.

Then suddenly the puzzle of the male calicos is solved. Many papers, some quite lengthy, have appeared on this topic since 1904, but as is often the case with seminal papers (like Watson and Crick's on the double helix and Mary Lyon's on X-inactivation), the announcement comes in a very short offering with an innocuous title. This one is called "Spontaneous Occurrence of Chromosome Abnormality in Cats" and appears in an August 1961 issue of *Science*. It's been two years since the discovery that human Klinefelters are XXY, and the authors, Thuline and Norby, offer evidence that the male calicos are their analogs. (It's interesting to note that they make this suggestion with no knowledge of X-inactivation; Mary Lyon's famous paper is published only four days after they finish writing theirs. It's also interesting that another researcher made the same suggestion independently in 1962, on the basis of knowledge of the Lyon hypothesis, but unaware of the paper described here.)

Thuline and Norby report on twelve cats with motley coats that they had hoped were all male calicos, although only one of them exhibited typical black and orange blotches. These are all tested for Barr bodies, but only two of the cats test positive: the black and orange guy and one with less obvious markings. The remaining ten tomcats who had been deemed possible candidates are disappointing and test negative. Their variegated coats are presumably the result of genes other than those at the orange locus.

Cells from the two cats with Barr bodies are cultured, but in these early days of such techniques the difficulties are sizeable. Only seven useful cells are obtained from the well-marked calico, only three from the other. All these cells, however, have 39 chromosomes and indicate that both cats are XXY. The calico is remarkable in that it has no internal reproductive organs of any kind (although it has a normal male phallus); the vet performing the exploratory surgery has never seen anything like it before. The other cat looks normal and has descended testicles, but no sperm are found within.

Practically simultaneous with this discovery is that of an XXY male mouse, which is also sterile. There are now three species — human, mouse, and cat — in which the same XXY chromosomal abnormality has been demonstrated, and it seems to cause sterility. Thuline and Norby are pleased to announce that cats can indeed be Klinefelters and should now join the ranks of mice as laboratory animals available for the study of this syndrome.

The cat people ride again

Whereas the early cat papers were concerned with finding the answer to the riddle of the calico cat, and thereby to some of the fundamental questions of genetics, many of the late cat papers have a more practical, human-related emphasis. Thuline and Norby had worked at a state school for the mentally retarded, and their work was supported by various associations for retarded children. When a reporter covering their work heard that male calicos might be helpful in understanding the causes of Klinefelter syndrome in humans, he put out a weekend plea for specimens in the local newspaper. On Monday morning the surprised switchboard operators at the school were inundated with calls from people offering cats and kittens of all descriptions — and that's where the twelve potential calicos came from.

Thuline's next paper, written in 1964 for the *Journal of Cat Genetics*, announces that he intends to study "the life history of male tortoiseshells" and closes with the words "We are convinced that these unusual animals will contribute to our understanding of a common human disease." Others point out that "these cats are among the largest nonhuman animals for which such abnormal chromosome patterns have been reported" and state that "the male tortoiseshell/calico cat is potentially the most useful mammalian model of chromosome aberration" available. Support comes not only from retarded children's groups, but also from the March of Dimes and the National Institutes of Health.

Serious scientists bother to write articles in popular cat lovers' magazines, explaining the genetics of the male calico and emphasizing the need for specimens for humanitarian research. They stress also the importance of gathering data on sterility, libido, facial ruff, intelligence and "quirks" in order to see how closely the cats follow the human Klinefelter model. Ads and articles appear in newspapers, vets are alerted, and a network of feelers goes out in many parts of the world. Still the male calicos are so rare (or their owners so unwilling to volunteer them) that by 1984, when the last of the calico papers is written, only 38 have been located for cytological examination.

Some of these cats, of course, turn out to exhibit more complicated chromosomal abnormalities than those of the two XXYs initially discovered. (Those two may not really have been so simple either. It's hard to know, given the tiny number of cells available for study.) In 1964, researchers at Oak Ridge National Laboratories report on a rare male calico with two separate cell lines: one is symbolized as 38XX (standard female), meaning that it consists of 18 pairs (36) plus XX; the other is symbolized as 57XXY, meaning that it consists of 18 *triples* (54) plus XXY. In this case, 38 cells are analyzed: 21 of them turn out to be standard female cells; the remaining cells have 3 chromosomes of each type instead of only 2. (Cells with triple chromosomes are usually lethal for most mammals, but further viable examples of this kind continue to turn up in the cat world as the search for male calicos gets under way.)

Such a double-triple structure is probably the result of the fusion of two embryos, one of which is triple already. The authors (now Chu, Thuline, and Norby) put forth a smorgasbord of possible explanations. The first, and simplest, proposes that the embryo with the triple chromosomes was formed by the simultaneous fertilization of a normal egg by two sperm, one X-bearing and the other Y-bearing, producing the 57XXY stucture. As if this weren't enough, this overly endowed embryo must then have fused with a normal female embryo to form the resulting chimera.

Musing on various possibilities, the authors point out that multiple cell lines could account for the fact that a few of the male calicos are actually fertile. This one is undoubtedly sterile, having no XY cell line to work with, but in his next paper, written in 1965, Norby suggests that "a pair of fused male twins could result in a male tortoise-shell which would have a normal appearing chromosome complement and most likely be fertile." (Perhaps I've maligned Ishihara and he could see perfectly well, even in 1956; this theory could explain his results if two male fetuses fused, one with a gene voting for orange, the other with a gene voting for non-orange.)

Sure enough, such a fertile tricolored cat is found in 1967. He's one of four stray kittens and turns out to be an XX/XY chimera, the result of the fusion of brother and sister embryos. Almost half of his cells (43 %) belong to the XY line, and these have made him a fertile male. This is the first reliable report of a male calico cat having no XXY cells at all.

Most of the cats studied have been of unknown or partially known parentage, but an XXY Himalayan with tortoiseshell markings at the extremities is found in 1971. This is a classy cat with a well-known pedigree, and the authors propose, with considerable justification, that he's the offspring of a fertile XXY male who produced XY sperm through non-disjunction. (This may have been the trick that Blue Boy used in 1956 to produce his eleven male calico offspring.)

Two similarly classy XXY cats, named Kohsoom Frosted Ice and Pyrford Ho Hum, are found in Australia in 1980. These pedigreed Burmese cats are both descended from the famous Kupro Cream Kirsch, the first cream Burmese cat imported into Australia from England. Both have blue-cream coats that are the dilute equivalent of calico. (When two dilute genes *dd* are present, black turns to blue, orange to cream; hence blue-creams are usually females only, just like calicos.) Even with the extensive pedigree information available, it's impossible to tell whether these closely related cats resulted from reluctance on the part of the mother or the father. In any event, the authors write a disclaimer, no doubt at the insistence of the breeders involved, stating that "there are no grounds for attaching any significance in relation to chromosomal abnormalities to the occurrence of this famous sire in both pedigrees."

The final paper also originates in Australia and is written in 1984. It reports on three pedigreed Burmese males with dilute tortoiseshell coats. Two are fertile and have normal XY karyotypes; the third is XXY and hence sterile. The XXY has evenly distributed coloration similar to that found in females, but the two XYs have very uneven distributions of black and orange hairs: One is 70–80 % orange; the other is about 95 % black, with a small orange patch on his forehead. With lots of data, from both karyotypes and pedigrees, the authors suggest that the situation of the two fertile males may be due to "gene instability" — a thought, they point out, similar to that expressed by C.C. Little when he talked about "a distinct mutation" way back in 1912.

They propose that the fertile males are making sperm of two kinds, some saying yes orange (*O*) and some saying no orange (*o*) as a result of genetic instability of the gene at the orange locus. Further, they suggest that the gonads of these cats "are mosaic in proportions that are in reasonable agreement with the ratios of color" in their coats. Evidence comes from their progeny: The mostly orange cat had thirty-nine daughters, thirty-seven of which received *O*, whereas only two received its non-orange counterpart *o*; the mostly black cat had nine daughters, only one of which received the orange gene *O*. Although chimerism (involving the fusion of two male embryos that disagree about the orange) might account for the first case, it can't be made to account for the second. Pedigree data show that his mother had no orange gene (she was *oo*) and was hence incapable of transmitting any orange color to her offspring.

Besides providing yet another theory, the authors provide a helpful table showing all the chromosomal complements of the thirty-eight male calicos that have been karyotyped and reported in the literature. The results are surprising:

> Less than a third are simple XXYs (although this was originally thought to be the answer to the puzzle).
>
> Slightly more than a third are complicated XXY mosaics (the most complex being 38XX/38XY/39XXY/40XXYY).
>
> About a third have no XXY component of any kind (16% appear to be XY, although no doubt some are XY/XY mosaics, whereas the remaining 18% are definitely XX/XY).
>
> Only 17% of all these animals are fertile. (Similarly, Mrs. Bisbee, reporting on only fourteen male calicos found during the period from 1904 to 1931, states that three, or 21%, were fertile).

But what about human Klinefelters? Have they benefited from this extensive research? Apparently not. Although there was never any suggestion that mental retardation could be cured, the hope was that the male calicos could provide more information on how chromosomal abnormalities occurred; this knowledge might then be used somehow for prevention. Dr. Norby, now retired, continues to breed cats on his own with such humanitarian goals in mind, but so far nothing useful has been discovered.

The suggestions made during the 1970s that records be kept of the relative size, intelligence, and personality traits of the male calicos to see what other correspondences with human Klinefelters they exhibited besides XXY-ness, were never carried out for lack of funding. But there's been no mention of mental retardation in the thirty-eight cats described, and George seems much smarter than average.

There's been no mention of long-leggedness, either, and I've been forced to admit, with some embarrassment, that I leapt happily to the conclusion that this was significant. Norby reports no known examples and suggests that George's long legs are due to a Manx gene lurking within (although his tail is perfectly normal). This fits well with the fact that George is the only cat I've ever seen who hops with his hind legs together like a rabbit — and I've just read that this is characteristic of the Manx.

So there seem to be no pressing open questions and no special ways in which the calicos can serve humankind. Still I wonder where George will fit into the chart and think it would be nice if he represented a type not yet found. I suppose future papers may yet appear announcing even more variations on the theme, and perhaps more theories about how they came to be that way, but the days in which the calicos presented some of the most interesting problems in genetics are no doubt gone forever.

The great cat chase

The 1956 paper by Komai and Ishihara was of particular interest to me, not because of its weak and erroneous conclusion about the crossing over of the X and the Y, but because of the large number of male calicos involved in the study. Whereas most researchers reported on a single animal (remember Lucifer and Blue Boy), or at most a few, often gathered from the files of others, these Japanese scientists claimed to be working with a sample size of sixty-five! (This turned out to be nearly twice as many male calicos as those located world-wide during the next thirty years.) To justify this unexpected number, they provide the following bit of folklore (in the midst of their technical paper in the *Journal of Heredity*) to explain how they came to have such an advantage over their fellow scientists of other nationalities.

> The Japanese people have a great interest in tortoiseshell male cats, because there is a current tradition that such a cat brings the owner good luck and security. Therefore the news of the birth of such a kitten often appears in local papers.

I was charmed and amused by this description and immediately envisioned Komai (or more likely his research assistant) sipping a cup of green tea while he scanned the morning papers in search of such announcements. What did they look like? Was there a standard form? I imagined a box, surrounded by stars, enclosing the Japanese version of the words "Unto us this day is born ..." but that didn't actually seem very likely. Was this still a "current tradition"? How could I find out? Another big search was clearly impending.

To my disappointment I learned that Komai was dead and, even worse, that his work had been discredited — his data were now thought questionable. This was sad, but no deterrent. Even if he'd invented some of the sixty-five cats or had chosen to see some of them as calico when they were merely tabby, he couldn't have made up the story about the birth announcements in the newspapers. They must have existed, and I wouldn't be satisfied until I'd located at least one example.

I called everyone I knew of Japanese descent: friends, my dentist, the man who repairs our generator; I called several Asiatic libraries and a Japanese information center, all to no avail. The Japanese acquaintances had never heard of such a practice, the libraries didn't hold newspapers, and there was a long pause at the end of the line at the information center. The young Japanese who was trying to cope with my unusual request said, "You know, Japan has changed a lot in the last decade. Semi-conductors, you know … ." It was clear that I was painting a picture of Japan that bore no resemblance to the one with which he was familiar. He was amused but polite and offered the addresses of the two major Cat Fancy organizations in Japan. (I wrote to both, but neither answered.)

Taking a cue from him, I used a high tech approach and asked friends to send messages of inquiry on international computer networks. That didn't work either. I then began to pester everyone I knew who was traveling to Japan, either on business or for pleasure, asking them to search for my "cat boxes," as I had started to call them. I must have asked dozens of people, but no announcements were forthcoming.

At last, by luck, I found a journalist just returning from Japan. He had many contacts in the world of the press, and one of his colleagues in Tokyo offered to participate in "The Great Cat Chase," as he bemusedly called it. This was high-class help and I was soon rewarded, not with the cat box I'd envisioned but with a long story from *Asahi Shimbun*, a large Tokyo daily whose name means "Rising Sun Newspaper." It was fairly current (1989), proving that male calicos are still worth a lot of column inches even in this sophisticated capital. And it provided a slightly different piece of folklore, which was central to the story.

It seems there was a baker, Mr. Yasuda, who lived in Funabashi-City. A stray cat decided to enjoy the warmth and comfort of his bakery, and she produced four kittens: two white, one black, and one calico (or "mi-ke" as they say in Japan). When Mr. Yasuda noticed that the mi-ke had testicles, he remembered from his childhood, when he had been evacuated to the country to escape American bombing during World War II, that fishermen in a Nagasaki fishing village cherished a male mi-ke as a guardian of their boats. This memory caused Mr. Yasuda to take the cat to a vet.

Figure 9. Sherlock Holmes of Funabashi-City.

The paper doesn't say whether the vet turned pale and swayed on his feet, but it does say that the director of the veterinary hospital was surprised because male mi-kes "are not supposed to exist." This doesn't daunt Mr. Yasuda, who has named his cat Holmes and reports that he loves to run through the house at night. "He is so cute. We will cherish him," says Mr. Yasuda. Not a very interesting story, but a great picture of Holmes is provided, who looks very much like George.

A picture is also provided of the head of the Primatology Department at Kyoto University, who is quoted as saying that we still don't know how male mi-kes occur and that they are all sterile. It's hard to believe that he actually said these things because he's probably acquainted with at least as much of the literature as I am. Oh well, I thought, what can you expect from articles written quickly for popular interest? Dealing with the press is known to be fraught with difficulty — errors are certain to occur.

Thinking about Mr. Yasuda and his beloved Holmes, I got to wondering, "Why Holmes?" The name seemed unsuitable, unlikely, and completely un-Japanese — perhaps something had been lost in the translation? I checked with my translator, who explained that the male mi-ke was named after Sherlock Holmes, a very popular figure in Japan. His popularity stems not only from the writings of Sir Arthur Conan Doyle but also from those of Ziro Akagawa, a Japanese writer who has produced eighteen very popular mystery stories for a largely teenage audience. The first of the series, written in 1978, involved a murder in which the victim's calico cat turned out to have a special talent for finding the murderer. The cat was thence dubbed "Mike-neko (calico cat) Holmes," after Sherlock Holmes and went on to star in the next seventeen volumes.

I was surprised to learn that this famous cat is female. If I were writing these stories, I would surely have chosen one of the rare males for the task of tracking down murderers. Perhaps Akagawa doesn't know they exist? Or perhaps the compelling reason was that females are thought by the Japanese to be better hunters than males? According to the translator, it's not unusual for the distinction between male and female names to be blurred in Japan, especially if the name is a foreign one, so Akagawa's choice of a male name for his heroine is not surprising — and Mr. Yasuda's choice of Holmes for his male mi-ke turns out to be particularly suitable.

Newspaper articles of this sort are no doubt what Komai was referring to, rather than the more formal birth announcements I'd imagined (in fact, according to the translator, there are no columns of birth announcements in Japanese newspapers); perhaps I'll eventually obtain more samples from the many feelers I've put out in all directions. But for now I'm content to admire the mysterious Japanese symbols, arranged in bold and artistic fashion around the picture of the male mi-ke, who is still capable of attracting so much attention.

Who knows where the truth lies?

In the beginning, when my reading revolved around cat magazines and textbooks, I often had occasion to wonder how much of what I was reading I should trust. For example, a well-known college text describes how "Mendel then tied little paper bags over the blossoms to prevent any wind-borne or insect-borne pollen from contacting the artificially fertilized plants." Having read this seemingly reasonable sentence, one carries around a mental image of these little (brown?) bags waving and bending in the breeze.

But Mendel himself writes that one of his major reasons for choosing *Pisum sativum* was that "the risk of adulteration by foreign pollen is ... very slight ... and can have no influence whatsoever on the overall result." Nonetheless, "a number of the potted plants were placed in a greenhouse during the flowering period ... to serve as controls ... against possible disturbance by insects." Did the author just make up these little bags? Did he copy the text from someone else, who had made it up earlier?

This same author tells us that "when Mendel published his results in 1866 ... it was known that most plant and animal cells have a distinct nucleus inside them, and that within the nucleus are rodlike structures called chromosomes." However, another well-known author reminds us that "it was historically impossible for Mendel to have been directly involved in cytogenetic problems. When he wrote his paper in 1865 chromosomes had not yet been discovered."

Besides all of these problems, there's the question of the data: how they were obtained, whether they're trustworthy, whether they have current validity. The cat people were happy to throw rocks at one another by blaming the sloppy record keeping of breeders. In 1952, Komai similarly discredits a Japanese "census of a cat population as to the coat colors and sexuality" (undertaken to determine whether the orange gene is sex-linked) by saying that "these census, however, were taken with the assistance of high school children, and they may not be quite accurate."

Examples of these and other problems abound and make it very difficult to form coherent historical images or to decide what further (mis?) information to pass along to the next set of innocent readers. I've done the best I could — but *caveat emptor*.

George Longlegs Rarity

Do people who own calicos communicate through some society? For a long time I thought not. The various cat fancy organizations are quick to point out that calico is not a breed — only a description of a color scheme found in many recognized breeds such as Japanese Bobtail, Maine Coon, Manx, Persian, and Rex. Hence calicos, they say, are not something to be registered by a society. Even the rare males are of no interest, since 83 % of them are sterile and the rest breed as though they were orange. Why would organizations concerned with perpetuating and refining a pedigreed line want anything to do with them?

But Judith Lindley disagrees. She thinks calico is a breed — a color breed — and wants it to be recognized. Besides being interested in calicos in general, she has a particular interest in males, two of which have been born into her household. In 1976 her calico cat Miss Kitty produced a male calico kitten who proceeded to father six litters. (The kittens, however, developed problems with their immune systems, and all but one died young.) Two years later Miss Kitty defied statistics by producing yet another male calico, but this one unfortunately was stolen before his fertility could be put to the test. These events caused Judith to pursue a quest similar to mine, and she encountered similar obstacles: unknowledgeable vets and no books or registries devoted to calicos. So she looked into the literature, contacted some geneticists, approached the cat fancy organizations, and decided to make a registry of her own.

It's unfortunate that Miss Kitty didn't do her thing earlier, when examples of male calicos were being eagerly sought all over the world. By 1978, when the Calico Cat Registry International (CCRI) was founded, interest in locating or studying these rare animals was definitely on the wane. And by 1991, when there were over 400 calicos registered with CCRI, twenty-five of which are males, no research was in progress. It's not clear what progress is being made on the recognition of color breeds either.

We register George anyway — only $2 for life — and go to visit Judith in order to study pictures of his counterparts. Most don't look like George at all: Many have large areas of white, and some have so little orange that the photo sports an arrow pointing to the few hairs that prove the existence of the orange gene. Since most of these cats have not been karyotyped, no one knows what their various genetic constitutions might be.

Registered cats often have complex, many-faceted names reflecting their parentage, coloration, and cattery — or perhaps just the desperation of the breeder in having to come up with yet one more monicker. I've come across pedigreed Burmese cats called Chindwin's Minon Twm, Lao's Teddi Wat of Yana, and Casa Gatos Da Foong; examples from other breeds include Roofspringer Milisande, Bourneside Shot Silk, Foxburrow Frivolous, Anchor Felicity, and Philander Carson — the list could go on and on. I wonder how you summon these cats. I just shout

"Here George" and he usually appears. I don't even know how to pronounce Chindwin's Minon Twm.

George has no pedigree, of course, but writing just George on the application form looked so lonely and bare; it didn't begin to fill up the ample space provided. So he's now registered as George Longlegs Rarity, an appropriate appellation, we thought, and the best that we could do. If you have a calico, male or female, that you'd like to register, just write or call Judith Lindley at CCRI, P.O Box 944, Morongo Valley, California 92256 (619) 363-6511. She'll be really glad to hear from you.

10 The Cat Is Out of the Bag

ON THE TWENTY-FIRST DAY OF AUGUST 1990, OUR THIRD GRANDCHILD IS born and George's chromosomes show up in our mailbox — rather pictures of them do, carefully arranged in neat rows to form his karyotype. Thus two important secrets are revealed in a single day: The child turns out to be a girl — 46XX as far as anyone knows; George turns out to be a mosaic — 38XY/39XXY as the Serology Lab tells us for sure.

Being a mosaic, George required two karyotypes instead of one. The first showed a standard tomcat pattern of 38 chromosomes — 18 pairs plus XY — in fact, those of Figure 6; the second showed a Klinefelter cat pattern of 39 chromosomes — 18 pairs plus XXY — and is presented in Figure 10.

So George isn't a new and wonderful type. He's just another boring old XY/XXY, five of which have already been reported in the literature. Actually, of course, XY/XXYs are quite rare, representing (before George) only 13% of the known cases. And George is rather special because he's a perfect fifty-fifty mosaic: Of the sixteen cells that were cultured, half were one kind, half the other. Thus, without our interference, George might well have been fertile since 50% of his cells are standard male ones, plenty enough to construct normal testes swarming with viable sperm. Of his five counterparts discovered earlier, only one was fertile, so most of them probably had cell populations of different proportions.

Figure 10. Klinefelter cat karyotype, 18 pairs of chromosomes + XXY.

Although this result is slightly disappointing, it's also quite satisfying since it appears to align itself so well with the observable facts: The XY cell line has produced George's obvious, well-descended sexuality, his considerable size, and his general fierceness with Oscar, Houdini, and other intruders; the XXY cell line has produced his beautiful black and orange blotches by providing two Xs that could disagree about the orange.

As I looked at the Klinefelter karyotype for the first time, I wondered idly which of the two Xs said yes orange and which one said no orange, but their inscrutable black shapes weren't telling. (They also weren't telling which X was inactivated because Barr bodies don't show in karyotypes, and all the chromosomes were failing to exhibit their characteristic banding because no one had asked them to. The slides containing George's white cells weren't stained for banding since this step would provide no further clues to the secret of George's genesis.)

And then it occurred to me suddenly that concentrating on this Klinefelter karyotype had perhaps caused me to think about the whole thing wrong: There's yet another X to consider, the X of the XY line, and perhaps *it* has produced the needed contrast. It all depends on the mechanics of George's origins — whether he's a run-of-the-mill mosaic or a rare chimera. And that's another thing his karyotypes aren't telling.

If George is a chimera, then it's entirely his parents' fault and he had nothing whatsoever to do with it. First one of them must have had sex chromosomes reluctant to part, causing an XXY zygote to be formed as shown in Figure 7; then his mother somehow produced an environment that caused this XXY embryo to fuse with a standard male XY one, producing an XY/XXY chimera. If this is what happened, then there's no knowing about the Xs. Those of the XXY line may be the same or different, but if they're the same as each other, then they must be different from the X of the XY line or George would not be calico.

It's more likely, however, that George is a mosaic. In this case both he and his parents are responsible, and my images about his Klinefelter karyotype are correct. Probably he started out as a standard XXY

Klinefelter cat, again as shown in Figure 7. But when George first started to practice division, he got it a little bit wrong: Instead of producing two XXY cells, identical with the original and with each other, he produced one XXY and one XY, having lost an X in the process. He appears to have an equal number of cells of both types, so he must have made this error early, perhaps on his very first try. Thereafter his division improved, and he proceeded to copy both cell lines faithfully forever after.

These words have a final ring and I realize, with both sadness and relief, that they're the last I have to say about this matter. George, now five, sleeps peacefully at my feet, shedding his parti-colored hairs all over the old Persian rug in the center of the Poonery. The sun is streaming though the glass door, making his tabby stripes and orange patches stand out vividly. Some of his secrets have been revealed, but others will remain George's forever.

Just as well, I think. Who would want to dispel all of his mystery? And just as well that he represents nothing new or original. Then people might be asking for just a little piece of his ear, or a snippet of some more private part, for further, more complex analysis. He and Max have their own lives to lead, and for that matter, so do I — it's high time for me to turn my attention elsewhere. But what a merry chase he's led me!

Afterword

It's been more than four years now since the Preface was written and the manuscript sent off on a lengthy odyssey of its own through the wilderness of modern publishing. A slight update seems in order.

Despite my forebodings, neither George nor Max has died. They seem quite unchanged at almost ten, though I've a few new crow's feet to my credit. Any wrinkles they might have are well concealed by shining fur, and they appear as agile as ever. Oscar too looks just the same, as does Houdini, now called Balzac - a more fitting name for his position as companion in an artist colony.

My uncle with Alzheimer's has died, succumbing at the age of 80 to the ravages of his terrible disease. Some have suggested that I should update the section regarding Alzheimer's in recognition of new genetic information that's been uncovered, but reading it over, I still think the library can wait. There'll be lots more false paths to follow, and no doubt years will pass before a real understanding, avoidance, or cure is possible.

My Japanese translator visited the Funabashi-City bakery in 1992 in the hope that she might encounter Holmes, but he was not at home. She had the wit, however, to exchange our 1991 Christmas card of George and Max for pictures of Holmes and his mother, who is also calico.

The Calico Cat Registry International now contains thirty-five entries for males, ten more than when we registered George Longlegs

Rarity in 1991. A thirty-sixth male has also come to my attention, this one advertised for sale for $10,000.

It seems a woman and her daughter brought three kittens to a pet shop; they thought one of them was a male calico, which they hoped might be worth from $50 to $100. When the store owner assured them the male wasn't really calico, they gave her all three kittens to sell for whatever run-of-the-mill kittens sell for. But soon the pet shop was advertising Sir Thomas of Corral for sale for $10,000, and the story hit the TV news all over California. It also hit the courts as the woman and her daughter, feeling cheated, sued the owner for $2500 and demanded the return of their kitten, now declared to be both rare and valuable.

The judge took quite a while to come up with his own version of the Judgment of Solomon. He explained that under normal circumstances the law is very clear: Someone who gives away a rock has no further claim upon it, even if it later turns out to be a diamond. "There is one difference here," he was quoted as saying. "To this date, we have not yet proved whether this cat is a diamond or a rock."

Having pointed out the crux of the contention, he had the wisdom to rule that Sir Thomas should remain in the custody of the pet shop, since he'd spent the last four months getting used to these surroundings. But the judge added the constraint that if the pet shop were to sell the cat for more than $100, then half the take must go to the mother and her daughter.

The pet store owner said she'd set the price at $10,000 because "that's the last price we heard of for a male calico," but I have no idea where she heard it. She also said that if he were priced too low, his rare genetic make-up could make him an attractive subject for scientific research and "I would never let that happen." Further questioning is difficult since this pet shop has since vanished from directory assistance.

The long drought has broken, to be replaced by floods and mudslides, but no major earthquakes have returned to shake our peaceful environs. Recently we were thrilled to see a baby mountain lion playing with its mother on the slope of a nearby hill. While the lioness stared at us benignly, without moving, her cub scampered for the cover of the brush along the meadow. Watching its progress, I realized that it was making straight for the spot where I'd been a few days earlier, all alone, crawling on my hands and knees in search of chanterelles; this patch is now off limits for the rest of the season.

We were not so thrilled by the report of a local ranch manager that he'd seen the mother, sleek with rain, prowling at night directly at the entrance to our drive. George and Max, however, remain unperturbed: They've learned the ways of the wild. They're streetwise. They're old and knowing.

Max prefers the kitchen, where the woodstove provides a cozy warmth. We've long since installed a sturdy latch that's technologically cat-proof but fails to take into account the weakness of the human psyche. So Max has learned that if he scratches persistently enough at that sliding door, he can persuade me to release its lock and open it.

George prefers the outdoors, even when it's cold and windy, and continues also to scratch persistently, leaving his muddy cat-prints on various closed glass doors he'd like to see opened, no matter what we say or do.

But on a day as grey and rainy as today, they've both decided to curl up here with me, nose to nose, inside the Poonery.

Poon Hill
February, 1996

Informal Glossary
(and mini-index)

The words listed here are relatively friendly, and so are their definitions, which aren't actually definitions in some cases but merely reminders from the text about their meaning. Before writing these short and sometimes flippant descriptions, however, I decided to consult some dictionaries and glossaries to see what others had to say. Here, for example, is the definition of *chromosome* from Webster's Third:

> one of the more or less rodlike chromatin-containing basophilic bodies constituting the genome and chiefly detectable in the mitotic or meiotic nucleus that are regarded as the seat of the genes, consist of one or more intimately associated chromatids functioning as a unit, and are relatively constant in number in the cells of any one kind of plant or animal.

Irene Elia, in her glossary for *The Female Animal*, describes chromosomes simply as:

> the thread-like bodies composed of DNA and protein, primarily in the nucleus of a cell.

This seemed much better, but I decided to write "a place where genes hang out" because, in the context of the material presented here, that seemed to be the chromosome's most useful attribute.

In deciding what words to include in this informal glossary, I selected for the most part those that I imagined a reader might stumble over, having put the narrative away for a while and lost the thread. Each entry is followed by the number of the page on which the term is defined. So if the short entries here are insufficient memory-joggers, one can follow the numbers for more detailed refreshment.

Barr body   a small dark blob indicating a highly condensed X-chromosome that's no longer in good working order (*see* X-inactivation) **131**

centromere   a belly-button-like object that ties two chromatids together (*see* dyad) **26**

chimera   a special type of mosaic that results from the fusion of two embryos **138**

chromatid   just a chromosome, more or less (*see* dyad) **26**

chromosome   a place where genes hang out (in the nucleus of a cell) **4**

chromosome complement   one pair of chromosomes of each type (for a particular organism) **34**

chromosome set   one chromosome of each type (for a particular organism) **27**

crossing over   an act occurring during meiosis in which chromatids intertwine their limbs and exchange commensurate genes **28**

dominant gene   a gene that can prevail against the wishes of its recessive counterpart **5**

dosage compensation   a mechanism for achieving equality between the sexes **146**

dyad   two chromatids tied together by a centromere; the result of a chromosome doubling itself during the initial stages of mitosis or meiosis **26**

factor   Mendel's term for what we now call a gene **81**

Felis catus   domestic cat — same as *Felis domestica*, but less euphonious **45**

Felis domestica domestic cat — same as *Felis catus,* but more euphonious **45**

freemartin a female mammal, usually a cow, that has been masculinized by sharing its mother's uterus with a male sibling **118**

gamete sex cell **65**

gene a sequence of DNA, resident on a chromosome, responsible for the production of a certain protein **5**

gene instability property of a flaky gene that manufactures different proteins at different times **172**

germ cell a cell that will be tranformed into either an egg cell or a sperm cell **27**

karyotype an organized picture of all the chromosomes in a single cell **36**

Klinefelter syndrome a condition found in 1 in 500 human males, characterized by sterility, feminine breast development, long legs, and mental retardation **135**

linkage the joint inheritance of several genes together **81**

location a gene's position on a chromosome **5**

locus location

Lyon Hypothesis one working X-chromosome is all a mammal is allowed (*see* Barr body, X-inactivation) **132**

meiosis the process by which sex cells are generated **27**

mitosis the process by which cells replicate themselves **24**

mosaic an organism that has two or more different cell lines **138**

mutant gene a gene that has suffered some random change from its "original" form (*see* wild-type gene) **50**

non-disjunction the inability of a pair of matching chromosomes to part when they should, either during mitosis or during the first or second divisions of meiosis (informal term: reluctance) **37**

polar body   a useless waste product of egg production **31**

polygenes   genes at different locations, all contributing to the same trait **134**

Punnett Square   a helpful box diagram indicating the genetic results of mating a male organism with a female one **18**

recessive gene   a gene that cannot prevail against its dominant counterpart **5**

reduction division   meiosis, which cuts the number of chromosomes in half **24**

reluctance   the inability of a pair of matching chromosomes to part when they should, either during mitosis or during the first or second divisions of meiosis (formal term: non-disjunction) **37**

segregation   the separation of chromosome pairs into two complete sets, each set including only one chromosome of each type **24**

set of chromosomes   *see* chromosome set

set of genes   all the different genes that can occur at a particular location (alleles) **5**

sex cell   either an egg cell or a sperm cell **27**

sex chromosome   a chromosome on which major sex-determining genes reside **6**

sex-limited trait   a trait appearing in a single sex only **113**

sex-linked gene   a gene that resides on a sex chromosome **11**

sex-reversed   having a chromosome constitution that is in conflict with observable sexuality **142**

spindle fibers   rope tows that help to pull the chromatids to opposite ends of the cell during mitosis and meiosis **26**

trait   the manifestation of a gene, or genes, at work **5**

translocation   the process by which a gene leaves home and takes up residence on a chromosome of a different type **142**

Turner syndrome   a condition found in 1 in 3500 human females, characterized by short stature, a webbed neck, and sterility **144**

variable expression   a property of some genes in which the degree of expression of a trait depends on how many genes of this type are at work **95**

wild-type gene   an "original" gene from which a mutant gene has arisen **50**

X-inactivation   the process by which mammals are restricted to having only one X-chromosome in good working order **132**

zygote   the single cell resulting from the fusion of an egg and a sperm **32**

Dateline

Important Events covered in this book.

1650 Harvey postulates the mammalian egg.

1677 van Leeuwenhoek sees the animalcules.

1827 Someone sees the mammalian egg.

1842 Nägeli sees a cell split in two and glimpses the chromosomes.

1851 Mendel enters the University of Vienna.

1865 Mendel presents his pea paper.

1873 Schneider sees the lining-up and poling-up phases of mitosis.

1883 Flemming sees the doubling-up phase of mitosis.

1885 Flemming sees the reduction division of meiosis.

1890s Weismann talks about "ids" being the bearers of hereditary traits.

1891 Henking describes "Doppelelement X," the X-chromosome.

1900 Mendel's pea paper is rediscovered; Bateson announces it in England.

1902 Sutton sees maternal and paternal chromosomes segregate in meiosis.

1902 McClung suggests that the X-chromosome is sex-determining.

1904 Doncaster writes the first serious paper about male calico cats.

1927 Mrs. Bisbee writes a huge review of the calico cat research.

1938 Dr. Turner writes about short, sterile women with webbed necks.

1942 Dr. Klinefelter writes about tall, sterile men with feminine breasts.

1949 Barr and Bertram see what came to be known as Barr bodies.

1956 Tjio and Levan find the correct human chromosome complement of 46.

1959 Klinefelter males are found to be XXY, Turner females X0.

1961 Lyon writes about X-inactivation.

1961 Thuline and Norby propose that male calicos are XXY.

1984 The most recent calico cat paper is written.

1986 George is born and the vets turn pale.

1988 The initial questions in this book are written down.

1990 George's karyotype reveals that he's a 50/50 mosaic, XY/XXY.

1991 The final answers to the questions in this book are written down.

1996 A brave publisher decides to chance the market.

References

I consulted many different kinds of materials during the construction of this book. There were books, professional journals, popular magazines, newspapers, television programs, fliers from cat shows, letters, electronic messages, and conversations with veterinarians, librarians, scientists, and cat lovers — a wealth of information provided through a variety of different media. A few of these materials have been mentioned specifically in the text; a great many more are provided in the references presented here.

These references are grouped into lists representing the following subject areas: history of genetics (concerning events, not specific to calico cats, that occurred before 1915), the early calico cat papers (1904–1949), the late calico cat papers (1950–1984), cats in general, cat genetics, genetics in general, and sex chromosomes and various sexual anomalies (Klinefelter and Turner syndromes, mosaics, and chimeras). Books are arranged alphabetically by author; papers are arranged by date of publication, and those widely considered seminal are marked with a star (*).

With the exception of the early and late calico papers, these lists are not intended to be comprehensive; nor are they to be considered pointers to the best available information in the field. They should be thought of more as itineraries, as lists of interesting way stations that I happened to visit as I navigated puzzling new terrain. Not all whistle stops are presented, by any means, but only those that contributed in some way to my delight, my understanding, or my prose. These itineraries, now carefully organized by topic, give no image of the haphazard paths of my many early explorations.

Most items listed contain references to a myriad of other items, so the reader is free, as I was, to choose further directions in which to roam. Serious travelers will need access to serious libraries, but the only other necessities are a hunger for knowledge, a sense of humor, and a resistance to claustrophobia. To all fellow travelers, serious or not, bon voyage and bon appétit!

History of Genetics
Books

Bateson, William.
Mendel's Principles of Heredity. London: Cambridge University Press, 1909.

Darwin, Charles.
The Origin of Species by Means of Natural Selection, or The Preservation of Favoured Races in the Struggle for Life (a reprint of the first edition). New York: Philosophical Library, 1951.

Darwin, Charles.
The Descent of Man, and Selection in Relation to Sex (second edition). New York: A.L.Burt, 1874

Darwin, Charles.
Animals and Plants under Domestication. New York: Appleton, 1890

Dunn, L.C. *A Short History of Genetics*. New York: McGraw-Hill, 1965.

Farley, John.
Gametes and Spores: Ideas about Sexual Reproduction 1750–1914. Baltimore: The Johns Hopkins University Press, 1982.

Iltis, Hugo.
Life of Mendel. London: George Allen & Unwin, 1932.

Irvine, William.
Apes, Angels, and Victorians. New York: Time Inc. Book Division, 1963.

Mivart, St. George Jackson.
The Cat. An Introduction to the Study of Backboned Animals, Especially Mammals. New York: Scribner, 1881.

Stern, Curt and Sherwood, Eva R., eds.
The Origin of Genetics: A Mendel Source Book. New York: W.H. Freeman, 1966.

Stubbe, Hans.
History of Genetics from Prehistoric Times to the Rediscovery of Mendel. Cambridge, MA: M.I.T. Press, 1972.

Sturtevant, A.H.
A History of Genetics. New York: Harper & Row, 1965.

Papers

1902* McClung, Charles E. "The Accessory Chromosome — Sex Determinant?" *Biological Bulletin,* vol. 3, no. 1–2, pp. 43–84.

1902 Sutton, Walter S. "On the Morphology of the Chromosome Group in *Brachystola Magna.*" *Biological Bulletin,* vol. 4, no. 1, pp. 24–39.

1902 Wilson, Edmund B. "Mendel's Principles of Heredity and the Maturation of the Germ-Cells." *Science,* vol. 16, no. 416 (December 19), pp. 991–993.

1903* Sutton, Walter S. "The Chromosomes in Heredity." *Biological Bulletin,* vol. 4, pp. 231–251.

1908 Stevens, Nettie M. "A Study of the Germ Cells of Certain Diptera, with Reference to the Heterochromosomes and the Phenomena of Synapsis." *Journal of Experimental Zoology,* vol. 5, no. 3, pp. 359–374.

1909 Wilson, Edmund B. "Recent Researches on the Determination and Heredity of Sex." *Science,* vol. 29, no. 732 (January 8), pp. 53–71.

1911 Johannsen, W. "The Genotype Conception of Heredity." *American Naturalist,* vol. 45, no. 531, (March), pp .129–159.

1911   Morgan, T.H. "An Attempt to Analyze the Constitution of the Chromosomes on the Basis of Sex-limited Inheritance in *Drosophila.*" *Journal of Experimental Zoology*, vol. 11, no .4 (November 20), pp. 365–414.

1913   Bridges, Calvin. "Non-Disjunction of the Sex Chromosomes of *Drosophila.*" *Journal of Experimental Zoology*, vol. 15, no. 4, pp. 587–606.

1916   Bateson, William. Review of "The Mechanism of Mendelian Heredity" by Morgan, Sturtevant, Muller, and Bridges. *Science*, vol. 44, no. 1137 (October 13), pp. 536–543.

1919   Roberts, Herbert F. "The Founders of the Art of Breeding." *Journal of Heredity*, vol. 10, no. 3, pp. 99–106.

1936   Fisher, Ronald Aylmer. "Has Mendel's Work Been Rediscovered?" *Annals of Science*, vol. 1, pp. 115–137. Also in Stern and Sherwood, *The Origin of Genetics: A Mendel Source Book* (New York: Freeman) 1966 pp. 133–172.

1950   Punnett, R.C. "Early Days of Genetics." *Heredity*, vol. 4, pt. 1 (April), pp. 1–10.

1952   Dawson, G.W.P., and Whitehouse, H.L.K. "The Use of the Term 'Gene.'" *Journal of Genetics*, vol. 50, no. 3 (January), pp. 396–398.

1954   Stomps, Th. J. "On the Rediscovery of Mendel's Work by Hugo de Vries." *Journal of Heredity*, vol. 45, pp. 293–294.

1957   Stern, Curt, and Walls, Gordon L. "The Cunier Pedigree of 'Color Blindness.'" *American Journal of Human Genetics*, vol. 9, no. 4 (December), pp. 249–273.

1964   Zirkle, Conway. "Some Oddities in the Delayed Discovery of Mendelism." *Journal of Heredity*, vol. 55, pp. 65–72.

The Early Calico Cat Papers

1904   Doncaster, L. "On the Inheritance of Tortoiseshell and Related Colours in Cats." *Proceeding of the Cambridge Philosophical Society*, vol. 13, pt. 1 (November), pp. 35–38.

1912   Little, C.C. "Preliminary Note on the Occurrence of a Sex-limited Character in Cats." *Science*, vol. 35, no. 907 (May 17), pp. 784–785.

1912   Doncaster, L. "Sex-limited Inheritance in Cats." *Science*, vol. 36, no. 918, p. 144.

1913   Doncaster, L. "On Sex-limited Inheritance in Cats, and its Bearing on the Sex-limited Transmission of Certain Human Abnormalities." *Journal of Genetics*, vol. 3, no. 1, pp. 11–23.

1914   Doncaster, L. "A Possible Connexion between Abnormal Sex-limited Transmission and Sterility." *Proceedings of the Cambridge Philosophical Society*, vol. 17, pt. 4, pp. 307–309.

1915   Cutler, D.W., and Doncaster, L. "On the Sterility of the Tortoiseshell Tom Cat." *Journal of Genetics*, vol. 5, no. 2 (December), pp. 65–72.

1919   Little, C.C. "Colour Inheritance in Cats, with Special Reference to the Colours Black, Yellow, and Tortoise-shell." *Journal of Genetics*, vol. 8, no. 4 (September), pp. 279–290.

1920   Doncaster, L. "The Tortoiseshell Tomcat — a Suggestion." *Journal of Genetics*, vol. 9, no. 4 (March), pp. 335–337.

1922   Bamber, Ruth C. (Mrs. Bisbee). "The Male Tortoiseshell Cat." *Journal of Genetics*, vol. 12, no. 2 (October), pp. 209–216.

1924   Tjebbes, K. "Crosses with Siamese Cats." *Journal of Genetics*, vol. 14, no. 3, pp. 355–367.

1927   Bamber, Ruth C. (Mrs. Bisbee), and Herdman, E. Catherine. "The Inheritance of Black, Yellow, and Tortoiseshell Coat-Colour in Cats." *Journal of Genetics*, vol. 18, no. 1 (March), pp. 87–97.

1927   Bamber, Ruth C. (Mrs. Bisbee). "Genetics of Domestic Cats." *Bibliographia Genetica*, vol. 3, 1927, pp. 1–83.

1927   Bamber, Ruth C. (Mrs. Bisbee), and Herdman, E. Catherine. "Dominant Black in Cats and its Bearing on the Question of the Tortoiseshell Males — a Criticism." *Journal of Genetics*, vol. 18, June, pp. 219–221.

1927   Tjebbes, K., and Wriedt, Chr. "Dominant Black in Cats and Tortoiseshell Males. A Reply." *Journal of Genetics*, vol. 19, no. 1, p. 131.

1928   Bissonnette, T.H. "Tortoiseshell Tomcats and Freemartins." *Journal of Heredity*, vol. 19, pp. 87–89.

1928   Bissonnette, T.H. "A Case of Potential Freemartins in Cats." *Anatomical Record*, vol. 40, no. 3, pp. 339–349.

1931   Bamber, Ruth C. (Mrs. Bisbee), and Herdman, E. Catherine. "The Incidence of Sterility Amongst Tortoiseshell Male Cats." *Journal of Genetics*, vol. 24, no. 3 (July), pp. 355–357.

1932   Bamber, Ruth C. (Mrs. Bisbee), and Herdman, E. Catherine. "A Report on the Progeny of a Tortoiseshell Male Cat, Together with a Discussion of His Gametic Constitution." *Journal of Genetics*, vol. 26, no. 1 (July), pp. 115–128.

1934   Wislocki, G.B., and Hamlett, G.W.D. "Remarks on Synchorial Litter Mates in a Cat." *Anatomical Record*, vol. 61, pp. 97–107.

1941   Koller, P.C. "The Genetical and Mechanical Properties of the Sex Chromosomes. VIII. The Cat (*Felis domestica*)." *Proceedings of the Royal Society of Edinburgh*, Section B (Biology), vol. 61, pp. 78–94.

The Late Calico Cat Papers

1952   Komai, Taku. "Incidence of the Genes for Coat Colors in Japanese Cats." *Annotationes Zoologicae Japonenses*, vol. 25, nos. 1–2, pp. 209–211.

1956   Ishihara, Takaaki. "Cytological Studies on Tortoiseshell Male Cats." *Cytologia*, vol. 21, 1956, pp. 391–398.

1956   Komai, Taku, and Ishihara, Takaaki. "On the Origin of the Male Tortoiseshell Cat." *Journal of Heredity*, vol. 47, pp. 287–291.

1956   Sprague, Lucian M., and Stormont, Clyde. "A Reanalysis of the Problem of the Male Tortoiseshell Cat." *Journal of Heredity*, vol. 47, pp. 237–240.

1957   Jude, A.C., and Searle, A.G. "A Fertile Tortoiseshell Tomcat." *Nature*, vol. 179, no. 4569 (May 25), pp. 1087–1088.

1961*   Thuline, H.C., and Norby, Darwin E. "Spontaneous Occurrence of Chromosome Abnormality in Cats." *Science*, vol. 134 (August 25), pp. 134–135.

1964   Chu, E.H.Y., Thuline, H.C., and Norby, D.E. "Triploid-Diploid Chimerism in a Male Tortoiseshell Cat." *Cytogenetics*, vol. 3, 1964, pp. 1–18.

1964   Thuline, H.C. "Male Tortoiseshells, Chimerism, and True Hermaphroditism." *Journal of Cat Genetics*, vol. 4, pp. 2–3.

1965   Norby, Darwin E. "Chromosome Abnormalities in Cats." *Animal Hospital*, vol. 1, pp. 263–265.

1967   Malouf, N., Benirschke, K., and Hoefnagel, D. "XX/XY Chimerism in a Tricolored Male Cat." *Cytogenetics*, vol. 6, pp. 228–241.

1970   Loughman, William D., Frye, Fredric L., and Condon, Thomas B. "XY/XXY Bone Marrow Mosaicism in Three Male Tricolor Cats." *American Journal of Veterinary Research*, vol. 31, no. 2 (February), pp. 307–314.

1971   Pyle, R.L., Patterson, D.F., Hare, W.C.D., Kelly, D.F. and Digiulio, T. "XXY Sex Chromosome Constitution in a Himalayan Cat with Tortoise-shell Points." *Journal of Heredity*, vol. 26, pp. 220–222.

1971   Gregson, N.M., and Ishmael, J. "Diploid-Triploid Chimerism in 3 Tortoiseshell Cats." *Research in Veterinary Science*, vol. 12, pp. 275–279.

1973   Centerwall, Willard R., and Benirschke, Kurt. "Male Tortoiseshell and Calico (T-C) Cats: Animal Models of Sex Chromosome Mosaics, Aneuploids, Polyploids, and Chimerics." *Journal of Heredity*, vol. 64, pp. 272–278.

1975   Centerwall, Willard R., and Benirschke, Kurt. "An Animal Model for the XXY Klinefelter's Syndrome in Man: Tortoiseshell and Calico Male Cats." *American Journal of Veterinary Research*, vol. 36, no. 9 (September), pp. 1275–1280.

1976   Centerwall, Willard R. "Calico and Tortie Males: A Riddle Solved." *Cats Magazine* (June), pp. 10, 46.

1980   Nicholas, F.W., Muir, P., and Toll, G.L. "An XXY Male Burmese Cat." *Journal of Heredity*, vol. 71, pp. 52–54.

1981   Hageltorn, Matts, and Gustavsson, Ingemar. "XXY-Trisomy Identified by Banding Techniques in a Male Tortoiseshell Cat." *Journal of Heredity*, vol. 72, pp. 132–134.

1981   Long, S.E., Gruffydd-Jones, T., and David, M. "Male tortoiseshell cats: an examination of testicular histology and chromosome complement." *Research in Veterinary Science*, vol. 31, pp. 274–280.

1984   Moran, C., Gillies, Chris B., and Nicholas, Frank W. "Fertile Male Tortoiseshell Cats: Mosaicism Due to Gene Instability?" *Journal of Heredity*, vol. 75, pp. 397–402.

1987   Switzer, Nancy Jo. "Simplified View of Color and Pattern Inheritance — Part XI: The Mystery of Red." *Cat World*, December, pp. EE12–EE13.

1987   Vanicek, Catherine. "A Few Basics on the Science of Genetics." *Cats Magazine* (August), pp. 16–18.

1987   Page, Susie. "The Male Calico." *Cats Magazine,* August, pp. 24–25.

Cats in General
Books

Aberconway, Lady Christabel. *A Dictionary of Cat Lovers: XV Century B.C. — XX Century A.D.* London: Michael Joseph, 1949.

Beadle, Muriel. *The Cat: History, Biology, and Behavior.* New York: Simon and Schuster, 1977.

Clutton-Brock, Juliet. *The British Museum Book of Cats.* British Museum Publications, 1988.

Fireman, Judy, ed. *Cat Catalog: The Ultimate Cat Book.* New York: Workman, 1976.

Nash, Mary. *While Mrs. Coverlet Was Away.* Boston: Little, Brown, 1958.

Mellen, Ida M. *The Science and Mystery of the Cat: Its Evolutionary Status, Antiquity, as a Pet, Body, Brain, Behavior, so-called "Occult Powers" and Its Effect on People.* New York: Scribner, 1949.

Mery, Fernand. *The Life, History and Magic of the Cat* (fifth edition). New York: Grosset & Dunlap, 1973.

Van Vechten, Carl. *The Tiger in the House* (fifth edition). New York: Knopf, 1960.

Weir, Harrison. *Our Cats and All About Them.* Boston: Houghton Mifflin, 1889.

Wright, Michael, and Walters, Sally, eds. *The Book of the Cat.* New York: Summit Books, 1980.

Papers

1881   Rope, G.T. "On the Colour and Disposition of Markings in the Domestic Cat." *Zoologist,* Third Series, vol. 5, no. 57 (September), pp. 353–357.

1917   Wright, Sewall. "Color Inheritance in Mammals: Results of Experimental Breeding Can Be Linked Up with Chemical Researches on Pigments — Coat Colors of All Mammals Classified as Due to Variations in Action of Two Enzymes." *Journal of Heredity,* vol. 8, no. 5 (May), pp. 224–235.

1917   "Ancestry of the Cat: Tabby an Animal of Mixed Blood — Egyptian Wild Cat Probably First Domesticated and Has Crossed with Other Cats in Many Lands to Which it Was Taken by the Phoenicians." *Journal of Heredity,* vol. 8, no. 9, pp. 397–398.

1918   Whiting, P.W. "Inheritance of Coat-Color in Cats." *Journal of Experimental Zoology*, vol. 25, no. 2 (April), pp. 539–569.

1919   Whiting, P.W. "Inheritance of White-Spotting and Other Color Characters in Cats." *American Naturalist*, vol. 53, no. 629 (November–December), pp. 473–482.

Public Television Programs

1990   "Cats." Nature, #507.
1991   "Cats: Caressing the Tiger." National Geographic Special, #1601.

Cat Genetics
Books

Jude, A.C. *Cat Genetics*. Neptune, NJ: T.F.H. Publications, 1977.

Robinson, Roy. *Genetics for Cat Breeders* (second edition). New York: Permagon Press, 1977.

Searle, A.G. *Comparative Genetics of Coat Colour in Mammals*. New York: Academic Press, 1968.

Papers

1934   Minouchi, O., and Ohta, T. "On the Chromosome Number and the Sex-chromosomes in the Germ-cells of Male and Female Cats." *Cytologia*, vol. 5, pp. 355–362.

1959   Robinson, Roy. "Genetics of the Domestic Cat." *Bibliographia Genetica*, vol. 18, pp. 273–362.

1963   Hsu, T.C., Rearden, Helen H., and Luquette, George F. "Karyological Studies of Nine Species of Felidae." *American Naturalist*, vol. 97, no. 895 (July–August), pp. 225–234.

1965   Hsu, T.C., and Rearden, Helen H. "Further Karyological Studies of the Felidae." *Chromosoma*, vol. 16, pp. 365–371.

1965   Jones, T.C. "San Juan Conference on Karyotype of Felidae." *Mammalian Chromosome Newsletter*, no. 15 (February), pp. 121–122.

1973   Wurster-Hill, D.H., and Gray, C.W. "Giemsa Banding Patterns in the Chromosomes of Twelve Species of Cats (Felidae)." *Cytogenetics and Cell Genetics*, vol. 12, pp. 377–397.

Genetics in General

Books

Benirschke, Kurt, ed. *Comparative Mammalian Cytogenetics*. New York: Springer-Verlag, 1969

Cooke, F., and Buckley, P.A. *Avian Genetics*. New York: Academic Press, 1987.

Crowder, Norman. *Introduction to Genetics*. New York: Doubleday, 1967.

Curtis, Helena. *Biology* (fourth edition). New York: Worth, 1983.

Dawkins, Richard. *The Selfish Gene*. New York: Oxford University Press, 1976.

Dawkins, Richard. *The Blind Watchmaker*. New York: Norton, 1987.

Elia, Irene. *The Female Animal*. New York: Henry Holt, 1987.

Eldridge, Franklin E. *Cytogenetics of Livestock*. New York: AVI Publishing Co., 1985.

Gonick, Larry, and Wheelis, Mark. *The Cartoon Guide to Genetics*. New York: Barnes & Noble, 1983.

Peters, J.A. *Classic Papers in Genetics*. Englewood Cliffs: Prentice-Hall, 1961.

Singer, Sam. *Human Genetics: An Introduction to the Principles of Heredity* (second edition). New York: W.H. Freeman, 1985.

Papers on Chromosomes and Their Evolution

1956*   Tjio, Joe Hin, and Levan, Albert. "The Chromosome Number of Man." *Hereditas*, vol. 42, pp. 1–6.

1958   Taylor, J.H. "The Duplication of Chromosomes." *Scientific American* (June), pp. 37–42.

1960   Stern, Curt. "Dosage Compensation. Development of a Concept and New Facts." *Canadian Journal of Genetics and Cytology,* vol. 2, pp. 105–118.

1961   Bearn, A.G., and German, J.L., III. "Chromosomes and Disease." *Scientific American,* November, pp. 66–76.

1968   Ohno, S., Wolf, U., and Atkin, N.B. "Evolution from Fish to Mammals by Gene Duplication." *Hereditas,* vol. 59, pp. 169–187.

1973   Ohno, Susumu. "Ancient Linkage Groups and Frozen Accidents." *Nature.* vol. 244, August 3, pp. 259–262.

Sex Chromosomes and Various Sexual Anomalies
Books

Bandmann, Hans-Jurgen, and Breit, Reinhardt, eds. *Klinefelter's Syndrome.* New York: Springer-Verlag, 1984.

> Klinefelter, H.F., Jr. "Background, Recognition and Description of the Syndrome." In *Klinefelter's Syndrome* (see above), pp. 1–7.

> Rieck, G.W. "XXY Syndrome in Domestic Animals: Homologues to Klinefelter's Syndrome in Man." In *Klinefelter's Syndrome* (see above), pp. 210–223.

> Zang, K.D. "Genetics and Cytogenetics of Klinefelter's Syndrome." In *Klinefelter's Syndrome* (see above), pp. 12–23.

McLaren, Anne. *Mammalian Chimaeras.* London: Cambridge University Press, 1976.

Mittwoch, Ursula. *Sex Chromosomes.* New York: Academic Press, 1967.

Ohno, Susumu. *Sex Chromosomes and Sex-Linked Genes.* New York: Springer-Verlag, 1967.

Ohno, Susumu. *Major Sex-Determining Genes.* New York: Springer-Verlag, 1979.

Ohno, Susumu. *Evolution by Gene Duplication.* London: George Allen & Unwin, 1970.

Stern, Curt. *Genetic Mosaics and Other Essays.* Cambridge, MA: Harvard University Press, 1968.

Papers on the X-Chromosome

1948   Hutt, F.B., Rickard, C.G., and Field, R.A. "Sex-linked Hemophilia in Dogs." *Journal of Heredity*, vol. 39, pp. 3–9.

1949*   Barr, M.L., and Bertram, E.G. "A Morphological Distinction Between Neurones of the Male and Female, and the Behaviour of the Nucleolar Satellite During Accelerated Nucleoprotein Synthesis." *Nature*, vol. 163, pp. 676–677.

1952   Graham, Margaret A., and Barr, Murray L. "A Sex Difference in the Morphology of Metabolic Nuclei in Somatic Cells of the Cat." *The Anatomical Record*, vol. 112, no. 4 (April), pp. 709–723.

1957   Austin, C.R., and Amoroso, E.C. "Sex Chromatin in Early Cat Embryos." *Experimental Cell Research*, vol. 13, pp. 419–421.

1961*   Lyon, Mary F. "Gene Action in the X-Chromosome of the Mouse (*Mus musculus* L.)." *Nature*, vol. 190, April 22, p. 372–373.

1962   Lyon, Mary F. "Sex Chromatin and Gene Action in the Mammalian X-Chromosome." *American Journal of Human Genetics*, vol. 14, no. 2 (June), pp. 135–148.

1962   Beutler, Ernest, Yeh, Mary, and Fairbanks, Virgil F. "The Normal Human Female as Mosaic of X-Chromosome Activity: Studies Using the Gene for G-6-PD-Deficiency as a Marker." *Proceedings of the National Academy of Sciences*, vol. 48, pp. 9–16.

1962   Whittaker, David L., Copeland, Donald L., and Graham, John B. "Linkage of Color Blindness to Hemophilias A and B." *American Journal of Human Genetics*, vol. 14, no. 2 (June), pp. 149–158.

1964   Ohno, S., Becak, W., and Becak, M.L. "X-Automosome Ratio and the Behavior Pattern of Individual X-Chromosomes in Placental Mammals." *Chromosoma*, vol. 15, pp. 14–30.

1965   Norby, D.E., and Thuline, H.C. "Gene Action in the X Chromosome of the Cat (*Felis Catus* L.)." *Cytogenetics*, vol. 4, pp. 240–244.

1967   Penrose, L.S. "Finger-Print Pattern and the Sex Chromosomes." *Lancet*, February 11, pp. 298–300.

1967   Grüneberg, Hans. "Sex-linked Genes in Man and the Lyon Hypothesis." *Annals of Human Genetics*, vol. 30, pp. 239–257.

1974   Lyon, Mary F. "Mechanisms and Evolutionary Origins of Variable X-Chromosome Activity in Mammals" (review lecture). *Proceedings of the Royal Society of London*, Series B, vol. 187, pp. 243–268.

1974   Ohno, S., Geller, L.N., and Kan, J. "The Analysis of Lyon's Hypothesis Through Preferential X-Activation." *Cell*, vol. 1, no. 4 (April), pp. 176–181.

Papers on the Y-Chromosome

1922   Castle, W.E. "The Y-chromosome Type of Sex-linked Inheritance in Man." *Science*, vol. 55, no. 1435 (June 30), pp. 703–4.

1957   Stern, Curt. "The Problem of Complete Y-Linkage in Man." *The American Journal of Human Genetics*, vol. 9, no. 3 (September), pp.147–165.

1960   Dronamraju, K.R. "Hypertrichosis of the Pinna of the Human Ear, Y-linked Pedigrees." *Journal of Genetics*, vol. 57, nos. 2–3, pp. 230–244.

1960   Gates, R.R. "Y-Chromosome Inheritance of Hairy Ears." *Science*, vol. 132, (July 15), p. 145.

1965   Dronamraju, K.R. "The Function of the Y-Chromosome in Man, Animals, and Plants." *Advances in Genetics*, vol. 13, pp. 227–310.

Papers on Sex Determination

1987   Page, D.C., *et al.* "The Sex-Determining Region of the Human Y Chromosome Encodes a Finger Protein.' *Cell*, vol. 51, (December 24), pp. 1091–1104.

1987   "Scientists Contradict 1987 Finding of Gene That Determines Sex." *Wall Street Journal*, (December 29), p. B4.

1988   Roberts, Leslie. "Zeroing in on the Sex Switch." *Science*, vol. 239, January 1, p. 21–23.

1989   Burgoyne, P.S. "Thumbs Down for Zinc Finger?" *Nature*, vol. 342, (December 21–28). pp. 860–862.

1990   McLaren, Anne. "What Makes a Man a Man?" *Nature*, vol. 346, (July 19), pp. 216–217.

1990   Page, D.C., et al. "Additional Deletion in Sex-determining Region of Human Y Chromosome Resolves Paradox of X,t(Y;22) Female." *Nature*, vol. 346, July 19, pp. 279–281.

1990   "Discovering What Little Boys Are Made Of." *Newsweek*, (July 30), p. 47.

Papers on Sex Chromosomes and Their Evolution

1961   Russell, Liane Brauch. "Genetics of Mammalian Sex Chromosomes." *Science*, vol. 133, no. 3467 (June 9), pp. 1795–1803.

1965   Ohno, Susumu. "A Phylogenetic View of the X-Chromosome in Man." *Annales de Genetique*, vol. 8, no. 1, pp. 3–8.

1991   Charlesworth, Brian. "The Evolution of Sex Chromosomes." *Science*, vol. 251, March 1, pp. 1030–1033.

Papers on Sexual Anomalies, Mosaics, and Chimeras

1938*   Turner, Henry H. "A Syndrome of Infantilism, Congenital Webbed Neck, and Cubitus Valgus." *Endocrinology*, vol. 23, no. 5 (November), pp. 566–574.

1942*   Klinefelter, H.F. Jr., Reifenstein, E.C. Jr., and Albright, F. "Syndrome Characterized by Gynecomastia, Aspermatogenesis without A-Leydigism, and Increased Excretion of Follicle-Stimulating Hormone." *Journal of Clinical Endocrinology*, vol. 2, no. 11 (November), pp. 615–627.

1959*   Jacobs, Patricia A., and Strong, J.A. "A Case of Human Intersexuality Having a Possible XXY Sex-Determining Mechanism." *Nature*, vol. 183, no. 4657 (January 31), pp. 302–303.

1959   Ford, C.E., and Jones, K.W. "A Sex-Chromosome Anomaly in a Case of Gonadal Dysgenesis (Turner's Syndrome)." *Lancet*, (April 4), pp. 711–713.

1960   Hannah-Alava, Aloha. "Genetic Mosaics." *Scientific American*, May, pp. 118–130.

1961   Russell, L.B., and Chu, E.H.Y. "An XXY Male in the Mouse." *Proceedings of the National Academy of Sciences*, vol. 47, pp. 571–575.

1966   Russell, Liane B., and Woodiel, Florence N. "A Spontaneous Mouse Chimera Formed from Separate Fertilization of Two Meiotic Products of Oogenesis." *Cytogenetics*, vol. 5, pp. 106–119.

1967   McFeely, R.A., Hare, W.C.D., and Biggers, J.D. "Chromosome Studies in 14 Cases of Intersex in Domestic Mammals." *Cytogenetics*, vol. 6, pp. 242–253.

1970   Lubs, H.A., and Ruddle, F.H. "Chromosomal Abnormalities in the Human Population: Estimation of Rates Based on New Haven Newborn Study." *Science*, vol. 169, no. 3944 (July 31), pp. 495–497.

1971   Cattanach, B.M., Pollard, C.E., and Hawkes, S.G. "Sex-Reversed Mice: XX and X0 Males." *Cytogenetics*, vol. 10, pp. 318–337.

1973   Hook, E.B. "Behavioral Implications of the Human XYY Genotype." *Science*, vol. 179, no. 4069 (January 12), pp. 139–150.

1974   Benirschke, K., Edwards, R., and Low, R.J. "Trisomy in a Feline Fetus." *American Journal of Veterinary Research*, vol. 35, no. 2 (February), pp. 257–259.

1974   Norby, D.E., Hegreberg, G.A., Thuline, H.C., and Findley, D. "An X0 Cat." *Cytogenetics and Cell Genetics*, vol. 13, pp. 448–453.

1975   Culliton, B.J. "XYY: Harvard Researcher under Fire Stops Newborn Screening." *Science*, vol. 188, no. 4195 (June 27), pp. 1284–1285.

1976   Witkin, Herman A., *et al.* "Criminality in XYY and XXY Men: The Elevated Crime Rate of XYY Males is not Related to Aggression. It May be Related to Low Intelligence." *Science*, vol. 193, no. 4253 (August 13), pp. 542–555.

1978   Ivett, J.L., Tice, R.R., and Bender, M.A. "Y two X's? An XXY Genotype in Chinese Hamster, *C. griseus.*" *Journal of Heredity*, vol. 69, pp. 128–129.

Question Index

The questions listed below are arranged in the order in which they appear in the text. Those above the dotted line were asked all at once, in a great rush of curiosity, on an August afternoon in 1988; those below the dotted line arose at various times during the following three years as a result of answers to the initial set. These questions are followed by page numbers indicating the beginning of a section wherein the answers can be found.

Index